Time Traveler

Time Traveler

A Scientist's Personal Mission to Make Time Travel a Reality

⌛ ⌛ ⌛

DR. RONALD L. MALLETT
WITH BRUCE HENDERSON

Thunder's Mouth Press
New York

Time Traveler:
A Scientist's Personal Mission to Make Time Travel a Reality

Published by
Thunder's Mouth Press
An Imprint of Avalon Publishing Group Inc.
245 West 17th Street, 11th Floor
New York, NY 10011

AVALON
publishing group incorporated

Library of Congress Cataloging-in-Publication Data is available.

ISBN-10: 1-56025-869-1
ISBN-13: 978-1-56025-869-8

Book design by Bettina Wilhelm

Printed in the United States of America
Distributed by Publishers Group West

Contents

"Flight by machines heavier than air is impractical . . . if not utterly impossible," stated one of the great scientists of the nineteenth century, Johns Hopkins professor of astronomy Simon Newcomb (1835–1909), who spent much of his career producing improved orbital tables for the moon and planets. Newcomb was convinced that a new metal or unknown force of nature would have to be discovered before man could consider taking to the skies. Even should a "power-machine" be invented that could lift a man off the ground, he predicted it would fall "a dead mass" to the ground and kill anyone aboard.

Simon Newcomb's comments about the impossibility of manned flight were published in 1902.

A year later, Orville and Wilbur Wright proved the expert wrong.

Prologue

June 25, 2002
Washington, D.C.

⧗

It was yet another hot and humid summer day in the nation's capital as I entered a large lecture hall on the sprawling urban campus of Howard University. Inside the air-conditioned hall, rows of seats rose upward from the podium, creating a theater-like setting that seemed apropos, as I was about to deliver the most important presentation of my life.

Since the age of eleven, I had told only a few confidants about my secret dream. Long fearing professional suicide, I had until recently held off revealing to colleagues at the University of Connecticut, where I am a professor of physics, my hope of turning one of man's favorite science-fiction fantasies into a scientific reality. Had I spoken of it earlier, I would not have received tenure.

The time had arrived. Before fifty or so of the world's leading physicists assembled here for the International Association for Relativistic Dynamics Third Biennial Conference, I was about to reveal in detail my plan for finally realizing my lifelong goal. It would not be enough for me to tell them my belief that this century will be the century of time travel just as the twentieth century was the century of air and space travel. No, this audience would want the nuts and bolts.

My work had been outed in the past year, which was no doubt the reason I had been invited to deliver a paper before this prestigious body, whose mission was "to facilitate the acquisition and dissemination of knowledge about research programs in classical and quantum relativistic dynamics of particles and fields." Yet, the fact that my research had been featured in publications as varied as *New Scientist*, the *Village Voice*, the *Boston Globe*, the *Wall Street Journal*, *Rolling Stone* magazine, and even *Pravda* (Moscow) would mean little at this august scientific gathering, and might even raise an eyebrow or two.

Sitting in the seats and peering down at me would be some of the heaviest hitters in my field of relativity physics: Bryce DeWitt, director of the Center for Relativity at the University of Texas at Austin and one of the founders of an early form of a quantum theory of gravity; Georgia Institute of Technology's David Finkelstein, who had made several important contributions, including a novel way of looking at black holes, and L. P. Horwitz, an influential Tel Aviv University professor who had hosted the last conference and had made a number of significant contributions to relativistic quantum mechanics.

Although my work is based squarely on Einstein's general theory of relativity—a solid foundation for any physicist to stand upon—I would be presenting some controversial results. This audience would expect to see the equations and solutions that led me to believe I had made a theoretical breakthrough that could lead to the design of the first working time machine.

For a few weeks, I had spent twelve to fifteen hours a day working on my calculations in order to be ready. I had transferred data to transparencies, which I intended to project during my talk. If I didn't get my math right, these experts would let me know, and would be none too tactful about it in the process. Should I make a mistake in a calculation and veer off course, I would be interrupted mid-speech and subjected to in-your-face critiques—"Professor, your equations are wrong"—rather than any gentle, instructive words of advice. That is the world of physics; we are, after all, scientists, not psychotherapists.

I was scheduled to deliver my talk at 10:00 AM that morning. Much to my chagrin, I saw on the schedule that Bryce DeWitt would be going right before me, presenting his paper on "The Everett Interpretation of Quantum Mechanics," about the many-worlds or parallel-worlds theory of the universe. That I would be following such a well-known star to the podium made me appreciate how the guy who came up to bat after Babe Ruth must have felt.

I knew I was in trouble right away when DeWitt started off by saying that a speaker needed only six transparencies to discuss his ideas, which, of course, was the exact number he had brought along. I looked down with dismay at my bulging folder, crammed with twenty-six transparencies. DeWitt went on to say that he always told his graduate students that they need not tell their audience everything they knew about a subject. I told myself to keep breathing.

DeWitt, a legendary and pioneering theoretical physicist, received his Ph.D. from Harvard in 1950. Tall, lean, and energetic, DeWitt had hiked in the Himalayas and Africa. During World War II he served as an aviator and after the war conducted research at the famed Institute for Advanced Study at Princeton, which was at that time also Einstein's academic home.

Elegant is an adjective often used by DeWitt's colleagues to describe his use of mathematics in physics, and it is certainly a compliment. The elegance in his work showed in the natural flow of his physical arguments, and the beauty was to be found in the pleasing symmetry of his equations. As a graduate student, I had learned that elegance and beauty in physics were nearly as important as whether an idea was correct. DeWitt's presentation about one of the strangest consequences of quantum mechanics—the possible existence of parallel universes—could play an important role in the possibility of time travel, so he had my rapt attention.

Quantum mechanics is the mechanics of sudden energy change. In quantum mechanics, energy cannot be gained or lost continuously, but only in fits and starts. In 1913, the Danish physicist Niels Bohr,[1] often called "the father of quantum mechanics," showed that the electron orbiting the proton in the hydrogen atom can change its orbit only by either gaining or losing a certain definite amount of energy—no more and no less. These definite, or discrete, amounts of energy are called quanta of energy.

In 1957, physicist Hugh Everett III,[2] then a recent graduate of Princeton, first applied quantum mechanics to the entire universe, which resulted in his many-worlds, or parallel-worlds, interpretation of quantum mechanics.

Quantum mechanics, in short, is a world of probability. In the ordinary, everyday world, when a pitcher throws a baseball, it is possible to describe exactly where the ball is and how it's moving. In the world of quantum mechanics, we can only say what will probably happen next, as we can't know exactly what an object is going to do.

In applying quantum mechanics to the whole universe, Everett found that whenever there is the possibility of more than one outcome for an event, there is a potential split in the universe. For example, suppose that at lunch you are trying to decide between a cheeseburger and a tuna sandwich. At the moment you make the decision, according to Everett, the universe splits into two parts. There is a universe in which you have chosen the cheeseburger, and there is also an equally real universe in which you have chosen the tuna sandwich. These new universes are parallel and separate. The you in the universe with the cheeseburger is not aware of the you in the separate universe with the tuna sandwich. Although this idea of a parallel universe seems incredible, it is completely consistent with the proven theory of quantum mechanics.

Like everyone in the audience, I was familiar with Everett's parallel universe theory. I had used the cheeseburger-tuna sandwich example many times to explain the theory to general audiences. And yet, I was enthralled by De Witt's masterful though highly-technical discussions of concepts like the "wave function of the universe." When DeWitt finished, I went to the podium. Having not forgotten his pronouncement that a speaker needed no more than six transparencies and did not have to tell the audience everything the speaker knew about a subject, I decided to face the situation straight up. "Just to let you know—I have no more than sixty transparencies. When I found out that Professor DeWitt was in the audience, I felt as if I was back in graduate school, and that I'd want to show him everything I know about the subject."

DeWitt and the audience laughed, and I caught my first deep breath.

I began by reviewing the tenets of Einstein's general relativity theory, which in this crowd of relativistic physicists was akin to preaching to the choir. Forthwith, I began to outline my own theories based on Einstein's work, projecting on a screen my transparencies with illustrations, equations, and final solutions, which I said showed that space and time could be manipulated in a whole new way that would lead to the possibility of time travel into the past.

A satisfying thought hit me—my dream and I had come a *long way.*

Then, I realized that pencils and pens had been put to paper in those rows of seats above me and that my esteemed colleagues were busy scribbling away as they began working through my calculations.

One

A Death in the Beginning

⧗

Time stopped for me in the middle of the night on May 22, 1955.

My parents, Boyd and Dorothy, had hours earlier celebrated their eleventh wedding anniversary. Mom was the prettiest of all the mothers, and I idolized my smart dad. They looked happy that evening, and it was clear they were very much in love.

My two brothers and I knew we were loved, too. My earliest memory is of a happy family outing to a neighborhood park.

That night, a Saturday, my parents had invited a dozen friends over to our apartment in a new complex of projects at 1455 Harrod Avenue in the Bronx. The house was filled with music and laughter, and my mother had prepared a turkey dinner with all the trimmings. The adults mixed bourbon cocktails, told stories, cracked jokes, and smoked cigarettes. My father, a renowned prankster and electronics whiz, had wired speakers all around the house—including a new installation in the bathroom, which dispensed music whenever the toilet seat was lifted. As it grew late, the kids were shooed away. Snuggled up in my bed, I must have fallen asleep feeling happy and secure.

My father had big plans, and our future seemed bright. We were to relocate in a few months to Long Island, where he planned to open a TV repair shop. A hard worker, Dad had been holding down two jobs—working for Sigma Electronics by day and repairing TV sets on nights and weekends. He was skilled, it seemed, at everything he attempted. He had helped wire the new United Nations building in Manhattan, and made TV repair calls on the likes of Jackie Cooper and Walter Matthau, both of whom, in appreciation of his work, gave him a signed photograph. My father built our first TV set, complete with a magnifying screen that made the picture larger. Not long after that, he delivered to my maternal grandparents in Pennsylvania one of their town's first television sets. My grandfather's favorite pastime soon became watching women's roller derby, and he howled in delight whenever a collision sent skaters flying off the track.

My father was a handsome, robust man with a soft baritone voice. He had a natural warmth with people, and possessed a gentle manner and keen curiosity. Although he put in long hours, he was never too tired to answer my questions about how things worked. When he was twelve, he had lost his father, at age forty-three, to silicosis, a lung disease brought on from breathing silica dust that results from the manufacture of bricks, a job my grandfather had done for most of his adult life. I would come to recognize that my father was deeply affected by his boyhood loss. Driven by a determination to rise above a menial labor existence for himself and his children, he also seemed troubled by the possibility of his own early death.

Mom and Dad grew up and went to school together in Claysburg, a rural hamlet in central Pennsylvania. Claysburg's black population resided predominately in town on Shanty Row, where Dad's widowed mother lived in a company house next to railroad tracks, and in an outlying area called The Field, where Mom's family had a small farm with chickens. My two sets of grandparents—Ira and Etta Mallett and William and Pinky Kimbrough—grew up in Mississippi and together ventured north in 1917 to escape racial injustice and find a better life; they ended up in Claysburg, where my grandfathers went to work making bricks at the "Brickyard," as did most of the local men.

Somehow, given the closeness of the families, it seemed right to everyone when my parents fell in love. Married just before my father went to war, my mother was pregnant by the time he shipped overseas. Dad served as a battlefield medic with a unit that took part in the first crossing of the Rhine by U.S. troops in early 1945. Later, Mom revealed that he was at times haunted by the memory of the suffering and death he had seen in combat, although it was not something I heard him discuss. A few weeks after crossing the Rhine, he became a father when I was born in Roaring Springs, Pennsylvania, on March 30, 1945.

After the war my father enrolled at a New York electronics school under the GI Bill, and worked as maintenance superintendent for our building in exchange for free rent. After learning his trade he always had a good job, which allowed Mom to stay home with our growing family. My brother Jason was born a year after me, followed three years later by Keith, and then Eve, eight years my junior.

Dad preached the importance of an education, and would withhold my twenty-five-cent weekly allowance until I passed his quizzes, such as testing my memory of the multiplication tables. Once, frustrated by too many wrong answers, he took me over to the living room window. Below was a new highway construction project, and I could clearly see a team of ditch diggers bending their backs to the arduous labor. "Is that what you want to do?" Dad asked. I told him no. "Then you better take multiplication and school more seriously."

One day he brought home a crystal radio set with earphones and helped me set it up in my room. By adjusting a small coil I could hear AM stations. I was intrigued with how the device could snatch signals out of the air from some distance away. Although we had a big console radio in the living room, my crystal set seemed more amazing because I had built it with my own hands and it was so small. Another memorable gift from my father was a gyroscope with a metal rotary wheel, and a string attached to a spindle. When I pulled the string, it spun the wheel, which in turn danced atop a little pedestal until the rotation stopped. As I watched, it was clear that the rotation kept the wheel from falling, *but why?* Dad talked about the energy of the spin and the axis of the rotation, explaining terms I had never before heard but which made sense to me given his easy-to-understand examples.

At this juncture of their lives, my parents had much to look forward to—four healthy children, my father's new business, life in the suburbs. Our previous Christmas had been joyful. Dad took on added work to have extra money, as he enjoyed seeing stacks of presents under the tree. It was important to him for us to believe in Santa Claus, but by then I began to figure that the jolly bearded white guy had some help from Dad. Encircling the tree was a train track, and a Lionel electric train rigged to stop and go on voice commands.

My father was not always the life of the party. He would some-times sit alone in near darkness, reading poems into his tape recorder with opera music playing in the background. At these times he seemed very sad, and I didn't understand why. Only in later years would I begin to piece together some possibilities as to what may have been troubling him at those times.

That Saturday night, after the anniversary party guests departed, Mom told me later, she suggested cleaning up in the morning because they were both tired. With church in the morning, Dad said he would rather do it before bed and got started. As they worked side by side, they discussed our upcoming vacation to Claysburg; Mom wanted to take the train or bus and Dad favored driving. They had recently bought their first car, and we had been enjoying Sunday rides.

I always looked forward to going to Claysburg and visiting my three surviving grandparents. We spent most of our summers there, with Dad joining us for his two-week vacation from work. My brothers and I ran through the fields and hills with our cousins, reveling in the open spaces. My father and maternal grandfather, Grandpap, got along well and enjoyed one another's company. I recall a familiar scene of my father relaxing in a lawn chair reading an electronics magazine. In my memory, Claysburg is always sunny, and in the middle of one of those lazy summers.

My parents carried their discussion that night into the bedroom. When Mom turned off the bedside light, she heard Dad sigh deeply. Thinking he was exasperated with her insistence about not driving, she said, "Okay, let's talk about it tomorrow." She nudged him playfully, and his head fell off the pillow like a sack of flour.

Awakening to my mother's soft crying, I got up to investigate.

When I stepped into the hallway, I heard her whimpering. "Oh, Boyd. Boyd." She was in the kitchen with the lights on. I heard a strange voice, and saw that a policeman was with her. Down the hall in the other direction was my parents' bedroom. If I turned left, I'd go to the kitchen; right, to their bedroom.

I went right, opened their door, and stepped into the dark room.

My father was under the covers. He wasn't moving, but he looked fine to me. *Was he asleep?* I went around the bed. My brothers had silently followed me into the room. Before I could touch Dad, a policeman appeared and ordered us out. He took us into the kitchen where Mom, dabbing her reddened eyes with a wadded tissue, sat at the table shaking her head sadly.

We three boys lined up in front of her. She took a deep breath and looked up at us. Although I cannot remember her exact words, she told us that Dad was dead. I remember feeling as if I was stuck in some kind of a bad dream I couldn't get out of. Everything after that, in fact, became very dreamlike, and my memory is filled with fleeting impressions—some dreadful, others a bit odd.

For some reason, Dad's body remained in the house until Monday. Apparently there was a delay in finding the doctor, who was required to sign the death certificate; Mom also remembers something about a citywide work slowdown by undertakers. So Dad stayed in his death bed. As if in slumber, he lay there with the covers undisturbed. In the hallway outside the bedroom stood a uniformed officer intent on keeping out visitors. Mother, however, insisted on going in a number of times, and the guard always relented.

The next thing I recall are my aunts and uncles at the house with plates of food, and being taken aside by one uncle and told how I was now the man of the house. I was ten years old, two years younger than Dad had been when he had lost his own father.

At the funeral home a few days later I stood at my father's open casket, still feeling as if everything was unreal. Mom had Dad dressed in his blue suit, and he looked handsome. He seemed to have fallen asleep without any pain or agony registering on his face. He seemed so *alive*—as if I could nudge him awake and say, "Hi, Dad. It's good to have you back."

Since the night my father died, I had been in shock, and numb—at times feeling this was not happening. I experienced the same disconnect at the funeral home. Then we took the long, slow ride to the veteran's cemetery, where he was given a military funeral, with a bugler blowing taps and seven ramrod soldiers firing their guns into the air three times each, loud reports that caused me to jump. Mom, dressed in black, sat holding the folded U.S. flag from the casket.

As I stood looking down into that horrible hole a coldness came over me that I would not soon shake. As if awakening suddenly, everything around me took on a chilling clarity. Emotionally confronting my father's death for the first time, I began to comprehend that I would never see or talk to him again.

Dad's final resting place was in Long Island—the only one of us who made it there—ending up in death, ironically, where we had planned to move and start a new life together, and where he was going to open his new business.

I began crying a quiet, sorrowful weeping that came from deep within.

When I saw my father into his grave, he was thirty-three years old.

Two

My Secret Mission

After we buried Dad, Mom went to work at a Manhattan restaurant on Forty-second Street making salads and sandwiches to be sold in vending machines. She had regular hours, and every morning on her way to work she would take my sister, Eve, to the free daycare center located in the projects. I was responsible for making sure that my brothers and I reached the school bus on time, and for housekeeping chores like folding laundry, dusting, and mopping floors. Sometimes I helped with dinner—peeling potatoes to have ready when Mom arrived, or cooking hamburgers and creamed corn for my brothers and myself.

One of my jobs was to take a cart and wait in line once a month at a depot in the projects where free food was handed out by a government agency. I would return with our family allotment of flour, sugar, cheese, and other basics. My mother's employer regularly sent her home with restaurant leftovers. Looking back, I'm not sure how she managed, keeping us healthy, fed, and clothed while dealing with the emotional weight of my father's death. Late at night, I would hear her crying in her room, or come upon her at the kitchen table sipping milk and whiskey and listening to sad music. Mom's loss was mine, and her anxiety was contagious. Like her, I had no idea how we would manage, let alone be happy again, without Dad.

As I went about my new routine, I was aware that something had happened inside of me. It was if I had shut down, and was just going through the motions. Once a naturally exuberant child with a gregarious personality, I became withdrawn and sullen. My father's absence left a void in my life that nothing seemed to fill. The how-and-why questions I had always put to him now went unasked. Schoolwork lost its meaning, as I knew there would be no more weekly quizzes that I needed to be sharp for. Just hearing the classical music he loved could bring me to tears. It was unbearable to think I would never be able to meet him at the end of the day at his subway stop. I was always so overjoyed to be in his presence, and thrilled when he let me carry his toolbox the rest of the way home. With his sudden and unexpected departure from my life, my childhood ended, and happiness was no longer my natural state. In a real way, a part of me was lost forever.

Mom began dating, and even at my young age I understood it was because she was terribly lonely. Among a series of men was a truck driver with whom she would get into terrible arguments. I recall us driving one night through a dark tunnel somewhere, and Mom and this guy arguing violently. He drove fast and erratically, and I cowered in the back seat, fearing an imminent crash. I was glad when he stopped coming around. I wanted Mom not to be so lonely and unhappy, but none of the men she dated were anything like my father.

Two summers after my father died we moved to Pennsylvania. Much later, Mom told me we did so because my maternal grandfather had insisted she return home. Concerned about her behavior with men, as well as the stress of raising four young children on her own, Grandpap wanted us all under his roof so he could keep an eye on her and so Grandma could help with childcare. By then, Grandpap had retired from the Brickyard and they had left Claysburg—forced out when a state highway was built through their property—and moved fifteen miles up the road to Altoona, where my mother's brother lived with his family. Compared to tiny Claysburg, Altoona—population 100,000—was a metropolis. However, I never felt as comfortable there as I did in the Bronx or visiting sleepy Claysburg.

Mom went to work as a cleaning lady at a dress shop, and left every morning looking as if she would be taking over the front office any day. Cleaning lady or not, she had her pride, and when Mom dressed up it was impossible not to notice her. Customers began to ask her advice about a blouse or a dress, and she was soon promoted, becoming one of Altoona's first black salesclerks. It was a lesson not lost on me. I observed that when one communicated an inner pride, it was often returned in the form of respect from others. While I was fortunate to have had a smart and ambitious father for the decade I did, I am still blessed to have a proud and strong mother.

The year after we moved in, Grandpa took ill with the same Brickyard malady that had killed my Dad's father: silicosis, the dreaded "white lung disease." I knew Grandpap was really sick when the televised roller derby held no interest for him. Mom and Grandma took care of him round the clock, and in the end he required an oxygen tent. After his death at seventy-eight, Mom felt it was time for us to have our own place. Using Dad's GI home loan, she was able to buy a modest house in an all-white neighborhood, and Grandma moved in with us. It turned out the house was affordable because it was located at the bottom of a hill right next to the railroad tracks that ran through Altoona, which served as a hub of rail traffic between Pittsburgh and New York. The entire house would shake on its foundation whenever a speeding train passed.

The move to Pennsylvania brought about something more drastic than simply a change in scenery. To that point in my life, being "colored"—the description of African-Americans popular at that time—did not have a negative meaning to me. In the Bronx, we lived in a predominately Jewish neighborhood, and I had never faced any hostility. As the only black member of a white and largely Jewish Boy Scout troop, I had never been treated any differently. And every summer when we went on vacation to Claysburg, my brothers and I played with our cousins and their white friends without incident. I had every reason to believe that things would be no different in Altoona.

Once settled in our new neighborhood, my brothers and I went exploring to meet some of the local kids. We spotted a group of four white boys playing nearby, and ran up to say hello. They looked at us, and one of them spat, "Niggers." I was stunned. I had never been called that horrible name—not in the Bronx or in Claysburg. Not anywhere. Something snapped inside me. Leaping forth without regard for the fact that I was outnumbered, I pounded the boy furiously with my fists until he said he was sorry. None of the other white boys made a move, and my brothers remained frozen like statues.

That ugly welcome to Altoona made me conscious of my race in a new and negative way, and because of my already precarious emotional state, it made me more depressed. I had a strong sense of being an outsider, and longed more than ever for my past life—when my father was alive and we were happy.

After the death of Grandpap, my grandmother went rapidly downhill. She began wandering aimlessly around town, and once took a train to Chicago "looking for my children," believing that Grandpap and their children were waiting there for her. When she became unmanageable, Mom was given no choice but to have her committed to the state hospital, as there were no affordable care facilities for patients with advanced senility—what would likely be diagnosed today as Alzheimer's. I went with my mother on Sundays to visit Grandma at the hospital where she was to live out her final years. Once in a while I could glimpse in her vacant eyes the loving woman I had known, but for the most part Grandma left us well before her death.

Abject unhappiness is my overriding memory of my Altoona years. Losing interest in school, I began not showing up at school. Sneaking back home after my mother left for work, I would climb through the basement window and spend the day alone in my room. Mom found out only when a truant officer came to our door. Though I was threatened with restrictions, I cared little since I didn't have any friends or activities anyway, so my truancy habit continued through the seventh and eighth grades. Still mourning the loss of my father, I became increasingly isolated and embittered. Not interested in games, sports, or socializing with my peers, I escaped into magazines, books, and movies, many of them fantasy and science fiction, which succeeded in taking me away from my reality. I indulged in endless daydreams about faraway, make-believe places complete with good-versus-evil battles and noble heroes—worlds far more enchanting than my own.

I had been captivated with the Knights of the Round Table since Dad took us to the 1954 movie *Prince Valiant*. I had since identified even more with the gallant knights and their noble cause, wishing I could right a terrible wrong (my father's death) and protect a damsel in distress (my mother). Before long, I would even find my own Holy Grail, which would lead me on a long quest.

My life-altering discovery came from an unlikely source. *Classics Illustrated*, published by the Gilberton Company, reprinted 167 titles in comic-book format (I probably read all but a dozen or so), summarizing and illustrating for young readers the plots of classic books. They cost fifteen cents, a sizable expenditure in light of my twenty-five-cent weekly allowance. I had started reading

them while still in the Bronx, and by the time we moved to Penn-
sylvania my collection included *Knights of the Round Table,* based
on Thomas Malory's classic, *The Death of Arthur*; Sir Walter Scott's
Ivanhoe; Charles Dickens' *A Tale of Two Cities*; Robert Louis
Stevenson's *Dr. Jekyll and Mr. Hyde*; Mary Shelley's *Frankenstein*;
Jules Verne's *20,000 Leagues Under the Sea*; Henry Wadsworth
Longfellow's *The Song of Hiawatha*; Homer's *Iliad,* and Shakespeare's
Julius Caesar, Hamlet, and *Macbeth.*

I vividly recall the moment when I spotted *Classics Illustrated*
No. 133 on display in the magazine rack of a drugstore not far from
our house. The cover illustration struck me like a bolt of lightening.
A solemn-faced man, surrounded by a futuristic scene, sat atop a
strange machine that looked to be a cross between a motorcycle and
a one-man spaceship. The machine had various hoops, tubes, and
wires projecting from it, and tßhe man was grasping two levers that
appeared to control the machine. The title at the top read:

THE
TIME
MACHINE
H. G. WELLS

Entranced, I opened the magazine. On the first page was the
same man, pictured smoking a pipe while intently working on the
machine in a room with greenhouse windows. Tools laid out on a
wooden platform included screwdrivers, wrenches, small torch,
oil can, screws, bolts, electrical cords. The man was tightening a
strip of white tape around some part of the machine.

Printed in the text box at the bottom of that first page was the
following:

Scientific people know very well that time is only a kind of space.
We can move forward and backward in time just as we can move

forward and backward in space. To prove this theory, I invented a machine to travel through time. If you pressed one lever, the machine went back into the past. If you pressed the other level, the machine glided forward into the future. With this machine, I set out to explore time.

I was dumbfounded. *"We can more forward and backward in time just as we can move forward and backward in space. . . ."* The most incredible and wonderful thing I had ever heard, these words filled my wounded heart with hope. *"Scientific people know very well that time is only a kind of space. . . ."* Experts *knew* this? Did this mean—could it *really* mean—I might be able to go back in time and warn my father to go to the doctor, slow down, take better care of himself? Things to prevent that frightful night from happening. Could I change his fate? And mine? Could I bring him back? I dropped three nickels on the counter and raced for home.

Alone in my room, I sat on my bed and began reading:

One summer evening toward the end of the nineteenth century, some friends of mine were gathered in my home in Richmond, England, as I entered.

"Good heavens, man! What happened to you?"

"Were you in a wreck?"

"I have just traveled through time. I don't mind telling you the story, though most of it will sound false to you. But it's true, every word of it. . . ."

The time traveler told of having made the last adjustments on his time machine. "I gave it a last tap, tried all the screws again, and put in a final drop of oil. I climbed up, sat in the saddle, and took the starting lever in my hand." He pushed the lever forward slightly, and within seconds noticed that the clock on the wall had moved ahead five hours. He took a deep breath, gripped the lever with both hands, "and went off into time." While his machine stayed in the same place, the time around him quickly advanced, and he went through rapidly-flashing scenes of what the future would hold.

He eventually stopped to see the future for himself. The surface beneath him suddenly became uneven, and the machine reeled over, flinging the time traveler onto the ground. He found himself in the midst of a forest during a rainstorm. Righting his machine and checking the instruments, he saw that he had traveled 800,000 years into the future. He walked around in the rain, coming upon a huge, forbidding structure with no windows or doors.

Deciding to return to his own time, he took his position in the machine and prepared to pull back the control lever when he heard voices. People speaking a "very sweet and musical language" approached him. He soon knew he had nothing to fear from these "simple and childlike" people, who called themselves Eloi, and joined them for a meal. When he returned to where he had left his machine, it was missing. He found tracks suggesting that his machine had been dragged into the nearby structure and solid doors closed behind it.

Soon thereafter, the time traveler saved an Eloi from drowning in a lake. Her name was Weena, and she was beautiful. In thanks, she gave him flowers of a variety he had never before seen. He placed two of them in his pocket to examine later. In the days that followed, the time traveler and Weena fought off the Morlocks, strange creatures that resembled human spiders, in a nighttime battle in a burning forest. Finding the doors to the structure open and his time machine a few feet inside, he started to drag it outside when the doors closed. Surrounded by Morlocks, he quickly activated the machine and pulled back on the control lever. He ended up back in his time at the opposite end of the room—the same direction and distance that the Morlocks had dragged the machine. The time traveler then walked into his parlor where his friends had gathered. "Incredible? Yes," he admitted to his friends. "Take it as a lie, if you wish. I hardly believe it myself. And yet . . ."

And yet, he had those two flowers from Weena still in his pocket.

Before leaving my bedroom, I read the story several more times.

Then I went into the basement and shut the door behind me. I knew exactly what I was looking for: my father's tools, last used by him and brought with us in our move from the Bronx. I opened the first box with competing degrees of exuberance and reverence. Carefully, I laid them out before me. Screwdrivers, wrenches of various sizes, screws, bolts, nuts, electrical wires. I found in other boxes pieces of radio receivers, television tubes, an oil can, soldering iron, and rolls of electrical tape.

Although it would take days for me to gather the other materials I needed, my plan had come together that first day. I had a secret mission that made me eager to wake up in the morning, and I did so with renewed purpose.

I tried to build the machine exactly as I saw it on the cover of *Classics Illustrated*, using television tubes, discarded pipes, and other junk. From watching my father work, I knew something about electronics, and he had always been patient about explaining what he was doing. I spent hours down in the basement each day, putting together the old tubes, wires, electric parts, and attaching some old car tires for landing gear. Finally, it seemed I had something. I plugged it in and pulled on some levers, but nothing happened.

I was disappointed but not discouraged. The time-travel story had mentioned "scientific people," and I concluded that I needed to learn science. That would allow me to put things together correctly, and build a working machine. But first, I wanted to read more about the time traveler and Weena. I went to the public library and took out Wells' original book. I needed a dictionary for even the first sentence: "The time traveler was expounding a recondite matter to us." I had to look up "expounding" and "recondite" just to get to the second sentence. But I did so, and finished the book, along the way learning that the time traveler's machine had taken him back before his own time, too. That scenario gave me renewed hope that my dream was indeed possible, as I did not wish to go into the future and meet the gentle Eloi and do battle with the dreaded Morlocks. No, I wanted to go in the opposite direction.

My quest from then on was to prepare myself so that one day I could design a machine that would take me back in time to before May 22, 1955.

I wanted to see my father again.

Three

Einstein and Me

After my discovery of H. G. Wells' *The Time Machine,* my hunger for time travel stories turned insatiable.

In the summer of 1957, I discovered a comics series, *Weird Science Fantasy.* Issue #25 was entitled "A Sound of Thunder," adapted from a 1952 story of the same name by science fiction writer Ray Bradbury. I found the story—now considered a time travel classic—truly captivating. Though I had no way of knowing it at the time, the inventive plot began to prepare me for future consideration about the perplexing paradoxes of time travel.

The Bradbury tale involves a private company, Time Safari, Inc., which specializes in sending hunters back millions of years to shoot a dinosaur. To prevent tampering with the past, they carefully trace the life span of a particular animal and find out how and when it is to die naturally, such as an aging triceratops becoming hopelessly stuck in a pit of tar. They then take a hunter back in time just prior to that event, and he is allowed to slay the animal a bit prematurely. After the natural course of events takes place, which the dinosaur would not have survived in any case, the hunters are careful to remove all their bullets and leave nothing else behind.

One day, a wealthy businessman named Eckles walks into the Time Safari office with a check for $10,000, the fee for a dinosaur hunt. Eckles wants to down a tyrannosaurus rex, the largest carnivorous dinosaur. He meets the experienced guide, Travis, and the two men idly discuss the previous day's election in which a liberal presidential candidate named Keith defeated an arch conservative, Lyman, whom they agree would have led the country into the "worst kind of dictatorship." Eckles is briefed on the strict rules of the hunt—no stepping off the path, having contact with any flora or fauna, altering the destiny of other creatures, or otherwise "touching this world of the past." The reason: "A time machine is a finicky business. Not knowing it, we might kill an important animal, a small bird, a roach, a flower, thus destroying an important link in a growing species." Eckles agrees to the rules, and Travis leads the hunting party to a metal machine that resembles a flying saucer, "a snaking and humming of wires and steel boxes . . . [with] an aurora that flickered now orange, now silver, now blue." With "the merest touch of a hand," they are soon transported back more than sixty million years before the election of President Keith.

Landing in the Cretaceous period, when dinosaurs roamed the earth, they track a particular tyrannosaurus rex. When they come face-to-face with the massive creature, Eckles panics and runs away. Other hunters bring down the dinosaur with several shots, and soon thereafter a massive tree limb falls atop the creature, which had been intended as its natural cause of death. The bullets are cut out of the carcass, and the party departs. Eckles is waiting for them back at the time machine. Transported back to their time, they find that many things have changed, including the results of the election, with President Lyman now set to pursue his "anti-Christ, anti-human, anti-intellectual" agenda. As an exhausted

Eckles sits down, caked mud falls from the soles of his shoes. There, stuck in the mud is the smashed remains of a golden butterfly, stepped on by Eckles when, in full flight, he veered off the path into the jungle. All the hunters realize, with horror, what they have wrought. "A small thing," Bradbury wrote, "that could upset balances and knock down a line of tiny dominoes, and then big dominoes, and then gigantic dominoes, all down the years across time."

At the library I located Bradbury's original story, read it, and ruminated. For the first time, I began to consider the delicacies of time travel, and realized there could be major undesired effects on the present as a result of something a time traveler inadvertently did in the past. By changing something as minor as the life span of a butterfly, the world's future had been altered.

As reading became both an escape and a passion, I was spending so much time at the public library that all the librarians knew me by name. They let me check out more books than the rules allowed, even titles not in general circulation. I was also making regular trips to the Salvation Army, where I could buy used paperbacks for ten cents. With no hesitation about spending my school lunch money for books, I became so thin that my mother took me to the doctor. As a result of my poor diet, I had developed secondary anemia. The doctor prescribed iron supplement, and told Mom to serve me liver at least once a week. When Mom realized how I had been reallocating my lunch money, she began sending me off to school with a bag lunch. Before long, I found an after-school job at a local barbershop—shining shoes and sweeping floors—to earn money to feed my book habit.

Not long after my failed attempt to build a time machine, Albert Einstein entered my life. Not in person, as he had died in April, 1955, ironically, only a few weeks before my father's death. I remember seeing the headline in the *New York Times,* and gathering at the time that he had been a great man.

It was in the Salvation Army one fall afternoon in 1957 that I happened upon a thin paperback published in 1948: *The Universe and Dr. Einstein,* by Lincoln Barnett. I recognized the name and image of Einstein, standing beside a life-size hourglass that had the sun and the earth in the top and the stars at the bottom. I bought the book, and it proved to be an understandable account of the structure of the universe as well as of Einstein's work. His breakthrough, I read, was to treat time as a fourth dimension. He said that the universe had four dimensions—three of space and one of time—which linked together to form space-time. I was elated to discover that Einstein had studied time, and I knew I would want to learn more about his work in order to achieve my goal.

The Universe and Dr. Einstein was to become the second most important book of my childhood. At that formative age, I saw little difference between science and science fiction. What I read in the latter genre seemed for the most part as plausible as what I read in the former. In fact, learning about Einstein's work made me even more appreciative of H. G. Wells. In 1895, when *The Time Machine* was published, a full decade before Einstein described a four-dimensional universe, the English novelist wrote about time as a fourth dimension. How could Wells have guessed?[3]

Diving into the Einstein book with great enthusiasm, I read about a famous experiment done by two American physicists, A. A. Michelson and E. W. Morley, in 1887. From what I could understand, they wanted to find out how the speed of light, which I read was 186,000 miles per second, was changed by the motion of the earth. That speed meant little to me until I read it was fast enough to circle the earth 10 times in one second. Now that was fast! I didn't grasp the details of the experiment at the time, but it seemed that Michelson and Morley wanted to show that as you moved toward a light beam, the speed of the light beam should be different from the speed that you measure as you're moving away from the beam. To me, this seemed to make sense. If I was running toward someone who was running toward me, they would be coming at me a lot faster than if I was running away from them. But to the surprise of Michelson, Morley, and seemingly everyone else in the scientific community, they proved that the speed of light didn't change no matter how fast you were moving toward it or away from it. How could that be? I wondered.

Curious and excited, I read on. The answer, I learned, came in 1905 from a young patent clerk named Albert Einstein. Einstein said that light is special, and that, unlike everything else in the world, light has the same speed whether you run toward it or away from it. The problem, he pointed out, was with the clocks that we use to measure the speed of light. According to Einstein, the speed of light doesn't change because the clocks that we use to measure it change. In other words, light is not affected by motion, but clocks are.

On page 59 of *The Universe and Dr. Einstein* I came across the formula that showed how time was affected by motion:

$$t' = \frac{t}{\sqrt{1-\left(v^2/c^2\right)}}$$

The formula was called the Lorentz transformation, developed by Dutch physicist Hendrik Lorentz.[4] This equation was expressed in elementary algebra, but might as well have been ancient hieroglyphics. Nonetheless, the Lorentz transformation became a sort of mantra to me. While I didn't understand much of its meaning, I did feel its power. I would often repeat to myself and idly scribble this equation for time as though it was some holy script, all the while not really knowing what was being said in the equation. Still, I did grasp something essential: Lorentz said time, represented by the letter t in his equation, could be affected by motion. Too, the fact that time could be represented by a symbol made it somehow less mysterious to me. Time was, in fact, a physical object that could be worked with and changed. Furthermore, Einstein's special theory of relativity, utilizing the Lorentz transformation, said that time slows down the faster you move. In other words, time slows down for a moving clock.

I did not fully comprehend all the articulations about the moving clock and the speed of light being the one constant in the universe, but I did keep coming back to time. To my eager twelve-year-old mind, Einstein's theory of special relativity served as an inspiration because it seemed to suggest that time travel was possible, which I took to mean that a time machine was also a possibility. Now, I just needed to better understand what Einstein was saying.

Turning into an avid collector of information and trivia about Einstein wherever I could find it, I soon learned that his interest in nature had been sparked as a boy by a compass his engineer father had given him. With warm memories of the crystal radio set and gyroscope given to me by my father, I felt my first real connection with the famous genius. I felt another bond when I read this inspiring Einstein quote: "Imagination is more important than knowledge. Knowledge is limited. Imagination encircles the world."

I was surprised to read that Einstein, as a young man, had to find a tutor and study hard to learn the sophisticated language of mathematics he would later need to present his revolutionary theories about time. I came across a statement Einstein had made that caused me to smile: "Do not worry about your problems with mathematics. I assure you mine are far greater." And this: "It's not that I'm so smart, it's just that I stay with problems longer."

In truth, all this talk about mathematics gave me pause.

Although I had an idea that—like Einstein—I would eventually need to learn a new language of mathematics to attain my goal, I hated arithmetic.

Four

In My Father's Footsteps

⌛

Understanding Einstein and the still mysterious Lorentz transformation seemed necessary if I was ever going to build a time machine. But I felt wholly inadequate, and no doubt with good reason. I was an inquisitive teenager who loved reading even difficult books, yet Einstein's work made my head hurt from trying to understand it.

A classic science fiction film, *Forbidden Planet*, suggested a plausible method for me to increase my brainpower. In the movie, a twenty-third-century space traveler visits a planet called Altair Four to find out why its colony of inhabitants had died off mysteriously. It turns out that the colonists had been destroyed by creatures formed by their "unconscious id." This was the first time I had seen the word *id*, and I looked it up at the library, finding out it was part of the human psyche that is the source of our psychic energy. Learning that the concept was developed by psychoanalyst Sigmund Freud, I went into the stacks to find his book, *The Ego and the Id*. Reading it, I became excited about the possibility of increasing the power of my brain so I could better understand the work of Einstein and other important scientists. Taking baby steps at first, which involved reading more and more challenging books, I eventually became a firm believer that people can build intellectual power just like physical muscles through exercise; the more one's intellect is stimulated, the stronger it becomes.

My mental powers were also expanded by two high school courses that caught every ounce of my imagination: electronics and algebra.

Motivated by my father's background as an electronic technician, I signed up for an electronics class in my junior year. A practical course designed for future electricians, it covered basic electrical circuitry theory and hands-on wiring. With recollections of my father showing me the inner workings of televisions, I now wondered if he had been preparing me to work in his business. I pictured the sign that would never be: "Mallett & Son TV Repair." My plan was to take my father's lead and become an electrical engineer, which seemed an ideal way to acquire the skills necessary to build a time machine.

In that first electronics class I learned about Ohm's law, and the man behind it. While teaching mathematics in high school in Germany, physicist Georg Simon Ohm discovered that at a constant temperature the flow of electric current (I) through a conductor is directly proportional to the potential difference, or voltage (V), and inversely proportional to the resistance (R). The equation he came up with is $I = V/R$. Although his theory was at first rejected, it soon came to be recognized as the basic law of electricity, and the physical unit measuring electrical resistance was subsequently named for Ohm.[5]

I learned that an electric current in a wire is essentially the flow of tiny particles called electrons. Electrons can be forced to move through the wire by connecting the wire to a battery, which is a chemical source of energy. The energy of a battery is measured by its voltage—named after the eighteenth-century Italian physicist,

Count Allesandro Volta (1745–1827), whose invention of the electric battery in 1800 provided the first continuous flow of current.

A 9-volt battery has more stored chemical energy than a 1.5-volt battery. The higher the stored energy of the battery, the greater its potential for moving electrons between different points of the wire. Hence, the voltage of a battery is also known as its potential difference. The greater the voltage, the larger the flow of electrons. If something is changed in the wire, such as adding a piece of wire made from a different material, then a resistance to the flow of the electrons occurs. This resistance is a friction that causes the added material—such as a filament—to heat up and glow. This is what happens when a lightbulb is connected to a wire that is attached to a battery.

My electronics teacher, John Bathgate, was a skilled and dedicated instructor. He required us to memorize all the permutations of the basic equation for Ohm's law. From the equation that described electric current flowing from a battery into a light bulb, I could see that if I put a larger battery (V) in the circuit there would be more electric current in the circuit (I), and if I increased the resistance (R) of the light bulb—such as by using one with a longer filament—it would decrease the current (I) in the circuit.

These equations were a revelation to me. I grasped how the symbols represented the elements in an electrical circuit. With this knowledge came my first insight into how mathematics connects to the physical world. Math would, from this time forward, never again seem abstract.

After going over the basics of house wiring, we moved further into electronics, learning how electrons are controlled to produce radio waves. When electrons are forced to move rapidly (oscillate) back and forth along a wire (antenna), radio waves are generated that move through empty space and are picked up by a receiver. The radio waves caused electrons in the receiver to oscillate. Those oscillations are converted into the vibrations of the diaphragm of a speaker. The sound we hear from the radio is due to the vibration of the air by the diaphragm of the speaker.

My cousin, Jones Kimbrough, was in my electronics class. A year older than me, Jones was also fascinated by electronics. We shared the notion that through electricity the world could be changed. We had to look only as far as radio and television, and wonder what other modern miracles could be shaped by electronics. In this context, I told him about my plans to build a time machine; in fact, he was the first person to whom I revealed my secret project. Jones did not laugh, and he told me he didn't think it was beyond the realm of possibility. He agreed that knowing electronics could be key to my success.

At the end of the course, we were required to present Mr. Bathgate with a binder of neatly written notes covering the entire curriculum. I stayed up several nights without sleep, rewriting my chaotic lecture notes. It was the hardest I had ever worked in school. I relished not only the learning experience but also the mental discipline. The work we did in the class with equations made me realize I needed to get over my math phobia and sign up for algebra, which I had been avoiding.

I took algebra my senior year, and to my shock found that it came naturally to me. I loved doing the equations, and would work and rework problems day and night for entertainment. Algebra seemed magical to me; numbers could be represented by letters, and then you could do *anything* with those symbols. I received nearly perfect grades and went to the top of the class. During one session, the teacher, Ethelyn Furrer, who had a way of making numbers come alive, talked about a form of higher mathematics that she said was even more interesting than algebra. Right then I decided that I would one day learn more about this intriguing subject called calculus.

As jazzed as I was about algebra, I had the opposite experience that year in physics, taught by a worn-out general science teacher who spent a lot of time reading aloud from the textbook. Writing on the chalkboard an equation such as the trajectory of a ball when it is tossed, he would show us the results without explaining the process, leaving out important details like what forces the ball went through from the time it was thrown until it landed. Although I passed the course, physics struck me as boring and somewhat incomprehensible. This disappointed me because Einstein was a physicist, and I continued to be captivated by anything I read about his life and work. For some years, this disconnect existed for me between the great man and physics, the same field that in future years would become for me a world of exploration and adventure, and in which I would find my own academic and scientific home.

My vastly different experiences in algebra and physics prove the value of inspiring teaching, and the positive impact on an impressionable student, as opposed to a lackadaisical teacher uninterested in lighting any fires. When I later became a class-room teacher myself, I tried not to forget that lesson.

With my newfound skill in algebra, I discovered that I could begin to comprehend the Lorenz transformation. Returning to the book *The Universe and Dr. Einstein* I could now see that, prior to the Lorenz transformation, time on someone else's clock (t') and time on my own clock (t) was always the same no matter how fast we moved with relation to each other. According to classical, pre-Einstein physics, time was an absolute, and regardless of our different speeds, the equation was always $t' = t$. This equation states that time is not changed by motion, and, therefore, time travel is not possible. But according to the exciting relativity physics of Einstein, if I'm moving very fast and someone else is standing still (or vice versa), then t' does not equal t. Because I am moving, the other person's time is not equal to my time. In relativity physics, time is changed by motion—and time travel *is* possible.

As I accumulated the combined knowledge of electronics and algebra, I was thrilled to begin to actually understand Einstein's work.

Things changed at home after my mother remarried in 1960. His name was Julius Oscar Williams, and we called him Bill. He was a decent man and good provider, which greatly improved our financial situation. Mom didn't have to work as many hours and was able to spend more time at home with the younger kids. And soon, there was a new one to care for: my baby sister, Anita.

Although at fifteen I was old enough to give Bill credit for taking on a widow with four children, we did not grow close. A high school dropout and nonreader, he had no interest in anything intellectual. In fact, he often badgered me whenever he saw my nose buried in a book. "What are you *doing?*" he would ask as if I was committing an indecent act. "Reading," I would answer without looking up. He believed that teenage boys should be interested in cars, girls, and hunting, the latter being his own passion. "There's something wrong with your oldest son," he would complain to my mother. "All he does is read and go to the library." Mom declined to get in the middle, and I kept doing what I liked best. In truth, something else came between Bill and me. I had elevated my father to such mythic proportions that I had no room in my life for a stepfather, who in any scenario would no doubt prove an inferior substitute.

The fire inside me that kept the memory of my father alive was further stoked by a chance encounter. It turned out that my mother kept all of Dad's reel audiotapes in the basement of our house. Discovering them one afternoon, I brought several of them upstairs. The first one I played was labeled *Rubaiyat of Omar Khayyam*. At first there was opera music, then I heard the deep, rich sound of my father's voice that I had nearly forgotten.

Awake! For morning in the bowl of night
  Has flung the stone that puts the stars to flight:

> And lo! The hunter of the east has caught
>   The sultan's turret in a noose of light.

I was stunned. This was one of the poems I remembered my father reading in the living room of our Bronx apartment, when he had seemed so sad and lonely.

> Lo! Some we loved, the loveliest and the best
>   That time and fate of all their vintage prest,
>   Have drunk their cup a round or two before,
>   And one by one crept silently to rest.

A sorrowful longing came from somewhere deep inside me. I wanted so much to see my father's kind face again. Then, as I listened, he recited the stanza that made me understand what this poem may have meant to him, and why it made him so sad.

> Ah, make the most of what we yet may spend,
>   Before we too into the dust descend;
>   Dust into dust, and under dust, to lie,
>   Sans wine, sans song, sans singer, and—sans end!

Once I looked up the definition of sans, the meaning of this stanza hit me. I remembered the preacher at my grandfather's funeral saying, "Ashes to ashes and dust to dust." I knew this poem had to be about the shortness of life. I was convinced that my father was reading this poem because he knew he was going to die soon. I listened to the tape over and over, unwilling to let go of this audible proof that I once had a father and knew happiness.

As I look back, I believe that this taped message from my dead father caused me to become obsessed with death—other people's as well as my own. I started to read the poems of Edgar Allan Poe, and could recite "The Raven" in its entirety at age fifteen. Death seemed a close companion at an age that I should have been feeling youthful immortality.

After reading, movies were my next favorite form of entertainment, and in that regard 1960 represented a special year with the release of *The Time Machine,* starring Rod Taylor as the time traveler. The film played at the State Theater in downtown Altoona for two weeks, and I saw it five times, settling each time in the fifth row center, with a box of buttered popcorn, a large Coke, and eyes riveted to the big screen. I was astounded by the time-travel special effects. Again this vivid story gave flight to my imagination, and made me more determined than ever to one day build my own time machine and see my father again. The promise of doing so seemed almost real enough to touch.

That same year a new TV show debuted that won my undying loyalty. *Twilight Zone* aired on Friday nights, and I always made a point of rushing home from the library, where I often went after dinner, stopping only to buy chocolate milk and donuts to snack on during the show. It was important not to miss the introduction by Rod Serling, who spoke over spine-tingling music set against a defused sky representing the mysteries of the universe, with images like a ticking clock and Einstein's $E = mc^2$ equation floating across it: "You're traveling through another dimension, a dimension not only of sight and sound but of mind; a journey into a wondrous land whose boundaries are that of imagination. That's the signpost up ahead—your next stop, the Twilight Zone!" Pour the milk and pass the donuts. *My* next stop . . .

A significant number of *Twilight Zone* stories involved time travel, which of course held me spellbound. In one classic episode, "The 7th Is Made Up of Phantoms," a trio of National Guardsmen conducting war exercises in a tank near the Little Bighorn encounters evidence that another battle is going on—the historic one that occurred in 1876. Their dilemma becomes whether to roll their powerful tank into the battle to save Custer and his men, and thereby alter history, or leave well enough alone and allow the massacre to happen. Compromising, they abandon the tank and enter the battle on foot, disappearing into the fray. Time lapse to present: their National Guard comrades find the deserted tank but no sign of the three men—until they check the names of the dead listed at the Custer Battlefield National Memorial not far away. I thought it interesting that the three time travelers had decided to join the battle and become participants in history while not attempting to change history, as the deployment of a modern—day tank at Little Bighorn certainly could have done.

From the world of science fiction, I developed an interest in computers, which were then primarily mainframes the size of a large room. They would soon start to shrink in size while computing power increased as a result of transistors replacing vacuum tubes. Compact computers were required for the U.S. space program, gearing up after the launch of the Soviet Union's first Sputnik satellite in 1957. Given all the space travel stories I devoured, I fully expected man would get into space, so the launch of Sputnik and the sudden inauguration of the Space Age did not come as a surprise. I do recall how upset people in Altoona were that the Soviets made it into space first and that their satellite could at times be seen streaking overhead, glinting against a starry backdrop.

Convinced that computers would be increasingly important in the years ahead, and hopeful that as artificial brains they might one day be an aid in building my time machine, I decided on the topic, "The Role of Electronic Computers in Our Future," for my senior oration in English. I did not mention anything about my time machine plans for fear of being ridiculed, but I did open with a kind of *Twilight Zone* scenario. I asked my classmates to imagine we were in the twenty-second century and that we had already landed on all the planets and were starting to reach out to the stars—all thanks to computers. I went on from there with some facts mixed in with a few fanciful predictions, at least some of which would come true in the decades ahead. My English teacher, Mrs. Rhodes, who had passed along to me and many of her students her love of Shakespeare, was kind enough to tell me that it was one of the most original senior orations she had ever heard.

Although I was coming into my own academically and feeling more confident, college was out of the question for financial reasons. I knew the only way I would ever be able to continue my education would be if I went into the military and then, when I got out, used the GI Bill to go to school, just as my father had done to acquire his electronics training.

I took an air force aptitude test and scored in the eightieth percentile. The recruiter said that my score combined with the fact that I had taken two electronics courses (the second one in my senior year involved radio receivers and repairing electronic circuits) would guarantee my entry into the air force's electronics school. Then he said something else—that the Strategic Air Command (SAC) was setting up a new system of computers nationwide for its command-and-control systems, and electronic computer technicians were needed. I signed on the dotted line.

The recruiter said he assumed that, like most boys just graduating high school, I would want to spend the summer at home and have a good time with my friends. He could arrange for me to start basic training in September.

"How soon can I leave?" I asked.

He laughed, thinking I was joking.

But of course, I was not.

"Sir, I want to leave as soon as you can schedule me to get out of here."

Two weeks later I was on a train leaving Altoona.

From a window seat I watched as we snaked through steep ridges cut into the majestic Allegheny Mountains. I could not know what the future held for me, but I was eager for it to start.

Five

A Project Under Development

⧖

Within minutes of stepping off a military bus at Lackland Air Force Base near San Antonio, Texas, I was in a group of other recruits being yelled at by a hard-nosed, ill-tempered training instructor nicknamed the Gray Fox. He pronounced us the worst bunch of "rainbows"—slang for recruits due to our multicolored civilian attire—he had ever seen. He said even our mothers could not save us now, and announced that our lives and assorted body parts belonged to him and the United States Air Force.

I began to think I had made a big mistake.

We ran everywhere and drilled incessantly in a stifling heat. A season in hell could not have been much hotter than that Texas summer. When it reached 120 degrees a red flag was raised, signaling there was to be no more running that day and everyone was to stay indoors. No flag went up when it was 110 or 115, and overheated recruits passing out became a daily occurrence.

Standard basic training lasted eight weeks, but because I was going on to a technical school I was let out after six weeks, and sent to my school for the final two weeks of basic before starting my electronics classes. Although I was happy to be leaving Texas, I wondered if I was going from the proverbial frying pan into the fire as I boarded a bus for Keesler Air Force Base in Biloxi, Mississippi. Keesler was supposed to be a top military electronics training center, and I was excited about the course work I would be doing there. On the other hand, growing up I had heard some very scary stories about Mississippi.

In 1912, some five years before the family migrated north, my mother's older sister, Lavinia, had been born in Kosciusco, Mississippi. Aunt Lavinia was a gentle woman who stood barely five feet tall. She was my godmother, and I loved visiting her and Uncle George in Claysburg during our summer vacations. As I got older, I wanted to know what life had been like when she was growing up in Mississippi. She saddened whenever she related some of the daily humiliations my grandparents suffered. She said that black people weren't allowed to go into certain stores to buy things. Aunt Lavinia remembered a time my grandmother took her shopping for flour. Grandma worked for a certain family picking cotton, and Aunt Lavinia remembered the store clerk shouting: "Hey, isn't you Miss so-and-so's nigger? Don't you know you're supposed to come in with her to get the flour?" Black people, more often than not, were referred to not by their given names, Aunt Lavinia told me, but as the "nigger" of the person they worked for.

Shortly after moving to Altoona in 1956, I heard from my cousins a terrifying story about a black youth from the North who was murdered while visiting relatives in Mississippi simply because he whistled at a white woman. I have since found that virtually every black male who grew up in the '50s and '60s heard this same story, and all were as terrified and angered by it as my cousins and I were.[6] Part of the lore of the African-American community, the chilling story served as a kind of coming-of-age for myself and many other young black males. "We best stay out of Mississippi," my Northern cousins and I had all agreed.

And here I was six years later on a bus heading that way. As we crossed into Mississippi, passing thickets of strange, flowered trees that someone identified as magnolias, I ruminated about the stories told by my aunt and cousins. I was returning to the very place my grandparents had fled to escape racial injustice. *Were times different now?* I wondered. *Had things changed in Mississippi?*

Biloxi is a small town located near the southeast corner of Mississippi on the Gulf Coast. The day was hot when we arrived at the base, and I felt my clothes sticking to me as soon as I stepped off the bus. Things got hectic right away, and for a while I had little time to think about those old stories.

When I received my first day pass to leave the base, I went into Biloxi with some classmates. The first thing I noticed were the signs, the likes of which I had never before seen. *"Whites Only." "No Colored."* Then there were the cold stares of the locals as I passed by with white servicemen. This was before the Civil Rights Bill and during Martin Luther King, Jr.'s active years; in so many ways, I realized, it was still the Jim Crow South of my grandparents' days. I understood very quickly why they had gone North in an effort to improve their lives, and why Grandpap had not permitted my mother to visit my father at a military base in Mississippi before he shipped overseas.

Hushed stories were told by air force personnel about blacks disappearing when they happened into the wrong sections of Biloxi. On base, we could go anywhere and be treated as equals due to federal law. But off base we couldn't go through the front door of a hamburger joint—"colored" were shooed to a back door or service window. I was terrified and angry at the threatening meanness of it all, of being judged outwardly by the color of my skin and not for my value as a human being. The bigotry that I had lashed out at shortly after we moved to Altoona now surrounded me.

When I returned from my first trip into town, my rage was palpable. Here I was in the uniform of my country, being asked to possibly die under Old Glory, while not being allowed to live in this country equally with other citizens. I made a vow, at that point, to remain on base for the duration of my training—six months of basic electronics and three months of advanced training. I resolved to spend my free time studying, reading or watching movies. With the exception of going home for Christmas, I kept my promise to remain on base.

The rigorous electronics training helped me to channel some of my anger and other emotions. The training was divided into twenty units called blocks. Thirteen blocks were devoted to basic electronics training and the remaining seven blocks consisted of an intensive study of electronic computer fundamentals. From my high school courses, I was already familiar with much of the basic electronics. The computer training, however, was fascinating and new to me. We were required to learn a special mathematical technique called Boolean algebra. This algebra aided our understanding of how computers processed data. Boolean algebra makes use of a binary code that is used by all computers. In binary code, only ones and zeroes are used to represent the usual numbering system. For example, the usual sequence $(0,1,2,3,4)$ would be represented by the binary sequence $(0,1,10,11,100)$. Computer circuits are quite simple devices that are either on (1) or off (0). Because of this simple circuit behavior, binary codes are a completely natural way for computers to deal with numbers.

In October 1962, my sense of personal isolation was temporarily dwarfed by the threat of all-out war. A few months after I arrived at Keesler, the nation was thrust into the nuclear face-off with the Soviet Union during the Cuban Missile Crisis. As we hadn't finished technical school yet, we speculated that they would give us rifles and send us down to Florida to protect the coastline because we weren't worth much else to the air force. It was a kind of gallows humor to relieve the tension, as we were certain that nuclear war was imminent and that mushroom clouds would soon materialize on the horizon. After the world avoided a nuclear holocaust, I went back to trying to survive Mississippi.

To ease my loneliness, I spent most of my waking hours in the base library when not in the classroom. My longtime interest in Einstein's work had turned into a hunger to learn more about the man behind the famous theories and equations. Browsing the stacks, I found *Einstein: Profile of the Man*, by Peter Michelmore. Settling in a well-lit quiet corner of the library, I opened the book and became immediately absorbed in Einstein's personal life.

The first thing about Einstein that surprised me was that he had children. In the author's note, Michelmore discussed having conducted an interview with Einstein's oldest son, Hans Albert. There had been no mention of Einstein having a family in anything I had previously read. The fact that the great genius had been a father made him seem all that more real to me.

Albert Einstein was born on March 14, 1879, in the town of Ulm, Germany, to a well-to-do Jewish family, and grew up in Munich. His engineer father, Hermann, moved the family to Milan, Italy, leaving Albert behind to finish high school in Germany. Albert was unhappy at the abandonment, and made life so unpleasant for his teachers that he was allowed to drop out. Albert immediately joined the family in Italy. Since he wanted to become a science teacher, he went back to school in Aarau, Switzerland, and attained his high school diploma. At seventeen he began studying at the esteemed Zurich Polytechnical School. Although his father urged him to learn a trade, such as electrical engineering, Albert was determined to study the world of science, and decided to specialize in mathematics and physics. After graduating from college, he married a classmate, Mileva Maric, a trained mathematician. Unable to find a teaching job immediately, he went to work at the Swiss patent office in 1902 in Bern, Switzerland, while attending the University of Zurich, where he would earn his Ph.D. four years later.

I responded to Einstein's fierce determination to take his own path in life. During trying times, he heartened himself with Emerson's line: "If a man plants himself indomitably on his instincts, the world will come round to him." Einstein was described as a "pleasant and conventional good-looking young man with a carefully trimmed black mustache and neatly brushed dark hair," which I had a hard time associating with the pictures I had seen of the elderly man with wild, funny-looking white hair. "Only the eyes dispelled an impression of the commonplace; the eyes were both brooding and brisk, restless with energy and intense in quick perception." The year before he received his doctorate in physics—1905—is known as Einstein's miracle year. He won his first international recognition that year with five published papers: the first on the photoelectric effect, in which he demonstrated the particle nature of light; the second on determining the size of molecules; the third on Brownian motion, demonstrating the existence of molecules; the fourth on his special theory of relativity, and the fifth paper a supplement that included his soon-to-be-famous mass and energy equation, $E = mc^2$. By the age of twenty-six, he was fully launched, with professorships and research positions from which to choose.

Einstein's early personal life was also eventful. In 1904, his first son, Hans Albert, was born. In 1910, while Einstein was a professor at Zurich University, a second son, Eduard, was born. Einstein's return to Germany in 1914 to become a professor at Berlin University seemed to cause problems between himself and Mileva. In 1919, Mileva and Einstein were divorced, and he soon married his cousin, Elsa. Einstein had a full life, which included playing the violin and sailing. I learned that he sometimes did his most important thinking while sailing his small boat on a lake. Even with everything that was happening in his personal life, Einstein was able to focus on his work. I learned that he considered his greatest triumph to be his general theory of relativity, sometimes called his "theory of gravity," which he published in 1916.

Michelmore mentioned a book of essays written by Einstein called *Ideas and Opinions*. Finding the book in the library, I was amazed by all the subjects in which Einstein was interested besides science. I could hardly believe it but there was an essay entitled, "Minorities." As I read what Einstein had written, I was deeply moved that he understood the problems that black people faced in this country. He wrote eloquently about the "tragedy" and "unfair treatment" of the "American Negro." This made me feel all the more that Einstein was truly a great human being as well as a great scientist.

"The power of our imagination," Einstein once said, "is greater than the power of our intellect." Of all the people in the world to have said those words they came from perhaps the greatest intellect of the twentieth century. *What did that say about Einstein's power of imagination? And, I wondered, what might he think of where my imagination was taking me? Would Einstein consider my dream to build a time machine wild and crazy—or imaginative and possible?*

After nine months, I finished the electronics and computer training. I was excited about what was going to happen next and especially about getting out of the South. As promised should I finish in the top ten percent of my class—which I did—upon graduation I was allowed to select a Strategic Air Command (SAC) base for assignment. I requested a location that was closest to Pennsylvania—and at the same time the farthest North.

I soon settled into my duties at Lockbourne Air Force Base outside of Columbus, Ohio. My job was to help maintain the computer system for SAC's refueling wing. I was responsible for a room-size, solid-state mainframe that monitored the distribution of the air tankers as well as the B-52 bombers to be refueled in-flight. Should the computer break down, which was seldom, I would bring online a backup system, then repair any hardware problem or send for a software specialist. It was easy duty, sitting at a desk watching the blinking green lights that meant everything was fine, and reacting on the rare occasions when a red light flashed. I volunteered for the graveyard shift—midnight to 8:00 AM—there were few supervisors around because they preferred working days. Enough of my fellow technicians were married and wanted to spend nights at home that I was able to work graveyard for the next two years. I sat alone in the control room through the wee hours, free to read and study to my heart's content as long as those green lights stayed on. The graveyard shift was great for catching up on my academic studies. The air force encouraged its personnel to take correspondence courses, and I signed up eagerly for different subjects. Since I had no social life, I had plenty of time to hit the books. When it was time to take a test, I reported to an office where I was given a timed examination with a proctor present. The answer sheets were mailed in, and I received the results back in the mail showing any questions I had gotten wrong.

Although it could be difficult taking advanced math as a correspondence course without a teacher present to explain equations, I enjoyed working and reworking the equations and was usually able to figure them out. In this way, I took algebra II and geometry and also an algebra-based course called solid-state devices—required by the air force as part of the continuing education for computer technicians—that described the elementary physics of semiconductors and transistors. I had received some training at Keesler in the area of solid-state devices such as transistors (the new SAC computers were all solid state, replacing older models with vacuum tubes), but this advanced course went into more detail about the behavior of the electron, the lightest electrically-charged subatomic particle.[7] I learned in this course that an electron behaves like a wave rather than like a particle. None of my earlier technical training had prepared me for this revelation. Since high school electronics, I had thought of an electron as a little billiard ball rolling merrily along through a vacuum tube creating sparks.

The authors of the correspondence course went on to say that an Austrian physicist named Erwin Schrodinger (1887–1961) had been the first to produce an equation to describe the wavelike behavior of the electron, which could pass through barriers and do all kinds of magical things. They pointed out that the mathematical training needed to understand the equation went far beyond the course, but that they would display the equation so it could be appreciated like a beautiful work of art, even though one might not understand what went into its creation:

$$\frac{\partial^2 \psi}{\partial x^2} + \frac{\partial^2 \psi}{\partial y^2} + \frac{\partial^2 \psi}{\partial z^2} + \frac{8\pi^2 m}{h^2}(E - V)\psi = 0$$

The authors were correct: I understood nothing about the equation, but was moved by its symmetrical beauty. I wrote the equation down in my own hand as if it would make more sense coming from my pencil. I loved the wavy lines, and the repetitiveness of the symbols. Inasmuch as the characteristics of an electron were responsible for this regal-looking statement, I was determined to devote my efforts to learning more about electrons in the hope that one day I would understand each symbol of Schrodinger's equation, for which he shared the 1933 Nobel Prize for Physics (with English physicist Paul Dirac) for "the discovery of new productive forms of atomic theory."

It was about then that I began to view myself as a project under development; I sensed that my intellectual wings were beginning to unfold. In preparing myself to be a knowledgeable, scientific person, I read constantly.

We had the option to eat our meals on base or receive a monthly food allotment and fend for ourselves (we could still dine in the mess hall but would be charged for meals). I opted for the allotment, and returned to my old ways of using food money to buy books. As I ran low on funds near the end of the month, my diet became somewhat limited: chocolate milk and potato chips.

I looked forward to my regular forays into town, which I dubbed my "Columbus Book Raids." One used bookstore became a favorite; it specialized in technical titles at low prices. There, I unearthed some remarkable books.

The Strange Story of the Quantum, by Banesh Hoffman, was non-mathematical, yet it was billed as being faithful to the basic concepts of quantum mechanics, which were described as the branch of mathematical physics that deals with atomic and subatomic systems. "The magnificent rise of the quantum to a dominant position in modern science and philosophy," Hoffman wrote, "is a story of drama and high adventure often well-nigh incredible." The book *did* read like an Arctic adventure, and I was mesmerized. As I read Hoffman's book, I realized that the people—Schrodinger and others—who wrote the mathematical equations to explain a world that had me completely captivated were theoretical physicists. For the first time I seriously began to wonder whether physics, rather than engineering, might be a better road for me to follow to gain the skills necessary to build a time machine.

Quantum Physics of Electronics, by Sumner N. Levine, was a comprehensive textbook for first-year graduate students. Although I was lost when I first cracked it on the graveyard shift, I was determined to keep at it until things became clearer. It would take a while. The preface recommended that the reader be familiar with vector calculus, elementary matrix algebra, and basic physics in order to fully understand the material covered in the book.

Unhappily, there were times when my nocturnal studies were cut short and my comfortable routine was interrupted. This happened on the infrequent occasions that I had to work either the day or swing shift. At these times, the ugly face of prejudice some-times made its appearance. One of the technicians was constantly preening himself. Someone must have told him he looked like Elvis Presley, and he thought he was hot stuff. One afternoon Elvis stopped at my desk and said, "Mallett, you look just like Sammy Davis, Jr." I had a strong sense from the way his words came out that it wasn't complimentary. Admittedly, I had lost weight on my meager diet, and could correctly be called skinny. I smiled, and said, "Yeah, and have you seen Sammy's wife?" I was referring, of course, to the blonde Swedish actress, May Britt. Elvis's eyes nar-rowed, and he emitted a forced laugh as he moved away. It felt good that I had gotten to him.

Ironically, I had recently read Sammy Davis, Jr.'s autobiogra-phy, *Yes I Can*, which had been a revelation of sorts. Here was a black man, highly successful, who lived life on his own terms. I identified with Sammy Davis's struggle against prejudice. When Davis was in the army during World War II, a group of southern recruits ganged up on him and covered him with white paint because he had been seen talking to a white female officer. In short order, Davis had used his talent as a performer to be trans-ferred into special services. From that point on, he realized that "my talent was the only thing that made me a little different from everybody else, and it was all that I could hope would shield me because I was different." Through Davis's story, I began to under-stand that I, too, might find protection from prejudice by utilizing my skills and talents in a way that would make it difficult for big-ots to get to me. This was a defining moment for me in terms of how to deal with prejudice. Davis's life story was also, I must admit, the beginning of a rebellious interest in white women. I began to develop elaborate fantasies about the type of woman that I might one day let in my life. (I had still never been kissed and wasn't dating.) Being able to communicate with her would be important, and also respecting one another.

Around that time I visited a cousin, Esther Reynolds, who was an X-ray technician in New York City. I always loved the stories she told about my father, whom she had adored. On this particular visit, I confided in her my loneliness, and admitted how shy I was around women. I bemoaned that I wasn't sure I would ever find a girlfriend. Esther told me not to worry, that when I got out of the air force and went to college, things would get better in that department. She said that the problem for me would probably be that I wouldn't be interested in a woman unless she was "as brilliant as Einstein and as beautiful as Marilyn Monroe." That instantly struck me as the perfect combination, indeed. Esther knew of my long interest in Einstein, but her innocent remark about Marilyn Monroe was the beginning of what would turn into another fixation for me. (To this day, the only photographs on the wall of my study at home are of Albert Einstein and Marilyn Monroe, which together never fail to raise the eyebrows of visitors. One is perfectly expected of a theoretical physicist; the other most certainly is not.)

I knew that Marilyn was considered the most beautiful woman in the world, and the object of most men's desires. I decided right then that I wanted to know more about her, and started reading everything I could find about her life and career. I learned that Norma Jean Baker had come from nothing to make herself something, and I admired her drive and determination. The rebellious side of me embraced the possibility that one day I might have a beautiful white woman like Marilyn Monroe or May Britt in my life. I found myself taking pleasure in wondering: *What would the crackers in Mississippi think then?*

Midway through my enlistment, my drive to get ahead was stronger than ever. If I was going to get to college, do well in my studies, change my life in a meaningful way, and achieve my goals, I knew I had to stay focused on my private studies of quantum theory and relativity during those long night shifts.

One title I liberated from a dusty shelf during a Columbus book raid was *Selected Papers on Quantum Electrodynamics,* edited by Julian Schwinger (1918–1994). I was fascinated by Schwinger's definition of quantum electrodynamics in the preface: "the theory of the quantum dynamical system in interaction with charged particles." I had only a vague idea what this statement meant, but I would keep this book next to me on various nightstands through the years, and it presently resides on a shelf in my study. For a time I gauged my growing knowledge of quantum mechanics by how well I was able to absorb Schwinger's preface. The book contained original papers by a number of well-known scientists, including "The Theory of the Positrons" by American physicist Richard Feynman (1918–1988). Unlike Schwinger, Feynman had a very accessible writing style. He explained his theory that a positron, which is the antiparticle of the electron, is actually an electron that is traveling *backward in time.* I was excited with the talk of traveling back in time as a scientific given. Due to the paper's easy-to-understand "Feynman diagrams"—a new graphical method he developed to represent the interactions of elementary particles—I was able to grasp the generalities of his theory right away, without comprehending all the mathematical calculations that he had no doubt used to reach his conclusions.[8]

A well-worn 1924 Dover paperback edition of *Principle of Relativity* had Einstein on the cover, so naturally I bought it. A collection of lectures and papers by Einstein and other renowned physicists, it traced the evolution of the theory of relativity. This book became a kind of holy scripture for me, and served as a guideline for where I wanted to go with my future studies in physics. I wanted to see if Einstein's fundamental paper on general relativity would shed some light on time travel. At the time, I didn't grasp much of the theory because it was a highly technical paper and involved a type of mathematics that was incomprehensible to me.

I showed Einstein's paper to my best friend in the air force, Paul Shattuck, who was also an enlisted computer technician. Paul was several years older than I, a college graduate, and the first true intellectual I had known. He had read all the classics, and was at ease discussing them, introducing me to notable philosophers such as Descartes and Kant. He had signed up to be an officer, but during flight training had become a born-again Christian and decided against being a fighter pilot. He was now serving the rest of his enlistment with a plan to enter the ministry when he got out.

"Paul, one day I'm going to be able to read this paper as easily as a comic book," I told him in all seriousness.

He laughed. The math was out of his league, too. "Well, good luck."

(Unfortunately, I have lost track of Paul over the years. Should I ever again meet up with my air force buddy, the first thing I would say to him is, "I did it, Paul. I can read it that easily now.")

Modern Science and Technology, edited by Robert Colborn, was a big book that cost nearly a week's worth of food money. Containing some eighty articles on various areas of science and technology, it included the article, "The Dynamics of Space-time," by John A. Wheeler and Seymour Tilson. They described in a non-technical way how space and time are warped by matter, and also that space and time were flexible, which was another exciting concept for my mind to absorb. Where gravitation in the long-accepted theory by English physicist and mathematician Isaac Newton (1642–1727) was a "force depending on matter for its existence, Einstein's giant step forward liberated gravitation from matter by describing it geometrically instead—as a curvature of space-time."

If there was any missing link in the reading I was doing in the air force, it came to me when I located a used copy of a 1949 edition of a book published upon the occasion of Einstein's seventieth birthday. *Albert Einstein: Philosopher-Scientist*, edited by Paul

Arthur Schilpp, contained an essay written that year by an Austrian mathematician and logician, Kurt Gödel (1906–1978), who in 1931 had made one of the most important discoveries of twentieth-century mathematics. Gödel's theorem states the impossibility of defining a complete system of mathematical rules that is also consistent, and as a consequence, math can never be placed on an entirely rigorous basis (i.e., there will always be some uncertainties). I read how Gödel was a close friend of Einstein's when they both were at the Institute for Advanced Study at Princeton, and this connection led Gödel into the field of general relativity theory. Also, Gödel had used general relativity to study cosmology, which I surmised had something to do with the universe. Then I came across this nugget in Gödel's essay: "*[I]t is possible in these worlds to travel into any region of the past, present, and future, and back again, exactly as it is possible in other worlds to travel to distant parts of space.*" Thrilled to read this statement, I underlined the passage and reread it until I could recite it from memory.

In the same book was this reaffirming comment from Einstein: "Kurt Gödel's essay constitutes, in my opinion, an important contribution to the general theory of relativity, especially to the analysis of the concept of time."

Although Einstein's general theory of relativity would not become clear to me until my later studies in college, I knew at that point it had to be key to unlocking the mysteries of time travel into the past. In the meantime, general relativity as described by Gödel and others was like a symphony to my ears; I enjoyed the music even though I couldn't read and understand the notes of the score or how the symphony was composed.

When I wasn't reading, my main sources of entertainment at Lockbourne were movies and TV. Since I still wasn't dating, I went to the movies alone and, as I had done since I was a youngster, sat up close with my popcorn and Coke. As I didn't report for work until midnight, I was also able to watch the TV programs that were shown every evening in the common room in the barracks.

In 1964, I became engrossed in a new science-fiction television series that captured my imagination as much as *Twilight Zone* had done when I was in high school. The new series was *The Outer Limits*. My attention was riveted the first time I heard the opening narration—reminiscent of Rod Serling's eerie introduction—spoken by the monotone "Control Voice" before each episode:

> There is nothing wrong with your television set. Do not attempt to adjust the picture. We are controlling transmission. If we wish to make it louder, we will bring up the volume. If we wish to make it softer, we will tune it to a whisper. We will control the horizontal. We will control the vertical. We can roll the image, make it flutter. We can change the focus to a soft blur or sharpen it to crystal clarity. For the next hour sit quietly and we will control all that you see and hear. We repeat: there is nothing wrong with your television set. You are about to participate in a great adventure. You are about to experience the awe and mystery which reaches from the inner mind to . . . *The Outer Limits*.

In many respects, *The Outer Limits* was the show that people like me, who had grown up watching the original—and since cancelled—*Twilight Zone,* had been waiting for ever since that cancellation. The new show had many of the same morality issues and twists of fate and faithfully stayed within the bounds of science fiction.

One evening, I saw a time travel episode that haunted me. "The Man Who Was Never Born" was one of the best of *The Outer Limits* time-travel stories. The episode starts off with an astronaut who has gone through a strange disturbance in space. He lands on a planet that is desolate, and encounters a grotesque creature that lives on the planet. The creature's name is Andro, played by Martin Landau. Andro tells the astronaut that he has somehow come through a time warp and has landed on the earth in the distant future. Andro explains that in the astronaut's time there was a biologist who had developed what he thought was an important cure for disease. It turned out that the so-called cure was a new plague. The human race was decimated by the plague and those who survived evolved into grotesque creatures like Andro. The astronaut suggests to Andro that, since he has come through a time warp, perhaps he could bring Andro back to his time, and together they could convince the scientist to stop his experiments. Andro agrees. When they go back through the time barrier, the astronaut is not able to make the transition and vanishes. Andro lands on the earth alone. Wandering around, he encounters a young woman. He's able to hypnotically suggest to her that he looks normal. Talking with her he is horrified to find out that he has arrived too soon. It turns out that the young woman is the future mother of the biologist who develops the plague. In the story, Andro tries to change the destiny of the young woman by wooing her from her fiancé. He convinces her to go back with him to the future. Unfortunately, Andro has sacrificed himself, because by changing her destiny he has changed his future and he is never born. As Andro and the young woman pass through the time barrier into the future, he disappears. End of story. "The Man Who Never Was Born" had a powerful effect on me. It made me think about ways that going back into my past could alter my life forever.

My final year in the air force brought me back down to earth. Transferred to administration, I worked as a clerk for six months. Just when I thought I was doing the most boring work possible, they turned me into a telephone operator sitting all day at a main switchboard. When it came time for a recruiter to interview me about reenlisting, I let him know that idea was so long dead it had already turned to ashes. Thank you, and good-bye.

Separated from the service in 1966, I went home to Altoona.

My stepfather, Bill, had lined up a full-time job for me that he thought I would be pleased with. He had pulled some strings and arranged for me to be the manager of a gas station in town. I must have looked at him like I had that recruiter about reenlistment. *Thanks, but no thanks.*

I had already been accepted into the physics program at Penn State.

Six

The Education of a Physicist

⧗

In Fall 1966, I started my new life as a college student at the Altoona campus of Pennsylvania State University, situated on the outskirts of town in a heavily forested area. Like other branch campuses of Penn State, Altoona offered a two-year curriculum; upon completion, students transferred to the main campus located in the town of State College to complete their four-year degrees. Although I wasn't crazy about the idea of returning to Altoona and living at home after being away four years in the air force, the situation would enable me to save money, with the GI Bill covering my tuition, books, and other expenses.

I had dreamed of the moment that I would step onto a college campus as a full-time student. I was eager to study *everything,* even philosophy, which, from my independent studies in the service, I knew might answer some of my nontechnical and unending questions about the nature of time.

My philosophy course was taught by a young, eager instructor named Charles Watkins. During class discussions, the two of us would frequently get into intense and lengthy debates about the meaning of the philosophy of Descartes and Kant. What I really wanted to do was get into a discussion about the nature of time, but I hadn't found many opportunities to do so. One day after class, the instructor took me aside and said he felt that I had studied beyond the level of the class, and had such understanding of the course material that he was willing to give me an A. He suggested that rather than attending class, I could have weekly meetings with him in his office to discuss those philosophical issues that interested me the most but which were not part of the curriculum. I think he wanted me out of the classroom so he could move through the course material with fewer interruptions. In any case, I was thrilled at the prospect of having one-on-one sessions with the instructor. In short order, after I described my interest in time, the instructor suggested I look up *The Confessions of St. Augustine* at the library.

Going to the Altoona library felt like a real homecoming; some of the librarians recognized me and welcomed me back to my old haunt. I found the book recommended by my instructor, and learned that Augustine (354–430) was a bishop of the early Christian church in North Africa. *Confessions* is actually comprised of thirteen separate books, each one a different mediation. The section that most interested me was entitled, "Time and Eternity." One chapter—"What Is Time?"—seemed as though it was written especially for me. Augustine asked, "What, then, is time? If no one asks me, I know: if I want to explain it to someone who does ask me, I do not know." At first Augustine's answer disappointed me, although I, too, felt I knew what time was until I had to define it in a precise way. As I read on, I found that Augustine continued to wrestle with the question of the "beginning of time." What exactly did the beginning of time mean—when the universe was formed or even earlier? I happily pondered this and other meditations by Augustine on the nature of time. Rather than be supplied with a bevy of answers, however, I found I had even more questions.

Although school was going well, I was having problems with living at home, as old patterns of behavior emerged. My stepfather, still clueless as to what I was trying to do with myself, kept after me to quit school and get a job. This only put more distance between us. It also made me ache all the more for the presence of my father, who I knew would have understood and encouraged my insatiable thirst for learning. To make matters worse, I was lonely and had no social life. These unhappy distractions soon began to affect my studies.

My smart and savvy German teacher, Ida Ficker, must have sensed something was wrong. She came to my rescue, insisting that I take part in the activities of the German club. Ms. Ficker, who regularly had students over to her home for dinner and socializing, also introduced me to a petite brunette named Marjorie Gey, who had graduated from Penn State a year earlier as a philosophy major and who had quit her social worker job in Philadelphia and returned home to help care for her mother, who had terminal cancer. Ms. Ficker asked Marjorie to give me a ride home (I had never bothered to get a driver's license) following a German club event. Marjorie was beautiful as well as brilliant, and once we were in the car she dived right into the technical aspects of the presentation we had heard that night from the German-American physicist Erwin Mueller, the inventor of the field ion microscope.[9]

Marjorie was interested in physics as well as philosophy. In her free time, she had been studying the highly abstruse work *Being and Time*, by the German philosopher Martin Heidegger (1889–1976). I told her I would like to know what Heidegger had to say about the nature of time. Marjorie struck a bargain with me. She would brief me on Heidegger's theories about time if I would explain the uncertainty principle of German physicist Werner Heisenberg (1901–1976), one of the founders of quantum mechanics. She had never completely understood the uncertainty principle in spite of having taken physics in college. We had ourselves a deal, and we began meeting after class for coffee.

I told Marjorie about my first reading of Heisenberg's theory in *The Strange Story of the Quantum* while in the air force. Quantum mechanics ruled the world of matter and energy, I explained, and the uncertainty principle is a fundamental aspect of quantum mechanics.

I recalled that in his book *The Feynman Lectures on Physics* Richard Feynman had given a classic example of why we need the uncertainty principle to explain certain phenomena. I decided to use Feynman's characteristically understandable explanation of the uncertainty principle in quantum mechanics.

I asked Marjorie to consider a simple hydrogen atom, which consists of a nucleus with one positively charged proton that is orbited by a negatively charged electron. The opposite charges of the proton and electron result in the attractive electrical force between them. Common sense dictates that as the electron orbits the proton it should lose energy and fall into the nucleus. This should happen so fast that all the matter in the universe should have collapsed long ago. Clearly, this did not happen. The reason why, I told Marjorie, is where quantum mechanics and the uncertainty principle come into play.

According to Heisenberg's uncertainty principle, it is not possible to know exactly both the position and motion of any atomic particle. This means even if you know the exact location you cannot know how it's moving. Conversely, if you know how it's moving, you cannot know exactly where it is.

I asked Marjorie to consider the situation of the electron in the hydrogen atom. As the electron tries to get close to the proton, the proton would know exactly where the electron is located. But, according to Heisenberg's uncertainty principle, the proton can't also know how the electron is moving.

The electron gets a little energy kick from the proton that moves the electron away from the proton. As the electron tries to get close again it gets another kick from the proton. If we average the distance the electron stays away from the proton due to the uncertainty principle, we find that the distance is exactly the size of the hydrogen atom. "In other words," I said, "without the uncertainty principle, the electron would crash into the proton and the hydrogen atom would collapse. So, you see, matter is stable because of quantum mechanics and the uncertainty principle."

Marjorie liked the Feynman example.

Now, it was my turn to listen to her.

She explained that Heidegger was concerned with a concept called *Dasein*, which is essentially another name for existence. In *Being and Time,* she went on, Heidegger addressed the problem of how we exist in time. The commonsense view of time is linear, and we think of time as a stream that flows past us. Yesterday is downstream and past us, while tomorrow is upstream and yet to reach us. According to Heidegger, time is not linear; the essence of our existence encounters all of time in one fell swoop. His concept of time takes into account all the multiple possibilities that potentially exist in the future.

My ongoing intellectual conversations with Marjorie helped me to focus again on my studies. As we continued to enjoy each other's company, she soon invited me home for dinner and to meet her parents.

Although it was not something Marjorie and I discussed, I knew the fact that she was white and I was black could cause possible problems for her parents. I wanted to believe that as a college physics major, an air force veteran, and a young man striving to better himself, I would be accepted for who I was. Marjorie's mother was friendly and I liked her immediately. But when Marjorie's father came home, the temperature in the room suddenly dropped to below freezing. Marjorie told me later that her father said she was never to bring me to the house again. That was fine by me.

Ironically, this incident brought us closer together. We continued to meet almost daily, and often talked for hours. One spring afternoon we were taking a ride to the neighboring town of Hollidayburg. Suddenly, somewhere on the outskirts of Altoona, Marjorie made a sharp turn onto a dirt road. Surprised, I asked where we were heading. Smiling, she said cryptically, "You'll see."

The road got steadily steeper for about a half mile. We pulled up in front of a sign that read "Chimney Rocks Park." On foot, I followed her up a rough hiking trail. As we approached a clearing, I could see an enormous rock formation that looked like a gigantic thumb jutting up out of the ground. The rock formation was near the edge of a cliff and as we got nearer a breathtaking view of the entire valley below opened up before us. Marjorie said this was her favorite spot in the world. The day was sunny, with a blue crystalline sky overhead. We sat next to each other, both spellbound by the beautiful setting.

Finally, I asked her what she wanted to do with her life.

"I want to learn more philosophy," she said.

She then asked me the same question.

"I'm going to find a way to build a time machine," I heard myself say, even though I had told only one other person—my cousin back in high school—about my obsession with time travel to the past. I went on to tell her about my father, and how I longed to see him once more. Marjorie commiserated with my loss, then asked a series of technical questions about time travel, some of which I could not answer but only touch on theoretical suppositions. Clearly this was one smart woman.

Marjorie and I seemed to share, at that time of our lives, a tragic sense of time and fate—as well as an unshakeable belief in the other person doing something important. Although we were never to become lovers, philosophy and physics became for us a kind of intellectual marriage.

After Marjorie's mother died, her father made it clear that she was no longer needed in Altoona. Not long after our visit to Chimney Rocks Park, Marjorie moved away to Chicago. I was heartbroken.

I retreated into myself and lost interest in my studies. As in the past when I needed to go to a different place to escape reality, I found solace in science fiction. The fall that I started at the Altoona campus marked the beginning of a phenomenon in the history of science fiction. It was the year that the TV series *Star Trek* first aired, and I quickly became immersed in the travels and adventures of Captain Kirk, Spock, and the rest of the crew of the starship *Enterprise*. An episode that aired near the end of the first season in 1967 had a particular resonance for me because it involved both the theme of time travel and lost love. This classic episode, "The City on the Edge of Forever," became my favorite of the *Star Trek* series. The crew of the *Enterprise* encounters distortions in time that are emanating from the surface of a planet. As they travel to the planet, the ship's doctor, McCoy, is accidentally injected with a drug that induces delusional behavior. McCoy beams down to the planet. When the crew arrives at the surface of the planet they find that the time distortions they had experienced in space were coming from a time portal built by some ancient civilization. A deranged McCoy runs through the portal. All of a sudden the crew finds that the starship *Enterprise* no longer exists. They realize that McCoy has changed something in the past that has changed their present. Kirk and Spock follow McCoy to the time portal and arrive in an American city in the 1930s. In searching for McCoy, Kirk meets and falls in love with a social worker named Edith Keeler, played by Joan Collins. As it turns out, Spock is able to determine that McCoy saved the life of Edith Keeler, who eventually leads a movement that delays America's involvement in World War II. Hitler is able to develop an atomic bomb and win the Second World War. In a sad and poignant scene, Kirk prevents McCoy from saving Edith Keeler from being hit by a car. Keeler dies and the crew returns to their present, where the starship *Enterprise* is now in existence again. The story brought home to me once again the potential intricacies of time travel, as well as the many unanswered questions that surrounded the subject.

Unable to concentrate on my studies, I left school after the fall semester of 1967 and moved to New York City with a woman I had met at Penn State who was fleeing an unhappy marriage. Dorothy Fry, who had plans to be a teacher, was nine years my senior. With dark brown hair, brown eyes, and olive complexion she looked Italian but came from German-Irish stock. Her most striking feature was her radiant smile, which everyone noticed right away.

Dorothy was the next person I told about wanting to build a time machine, and why. We were sitting in the hall of an Altoona campus building, two or three months after we met, as I related my story. Dorothy listened attentively as I explained with some emotion how I had dreamed since I was a young boy of going back to see my father and somehow preventing his fatal heart attack. "All of this emphasis on schooling," I said, "is about that dream. And it's why my education is centered on learning physics and mathematics." When I was finished, she looked at me with her expressive eyes, and said she had no doubt that I would one day succeed in building a time machine. At that moment, I leaned over and kissed her for the first time.

As first, Dorothy and I both were mainly interested in the other's friendship and having someone intelligent to talk to. Soon, it grew into much more. I found her easy to be with, and always felt a sense of calm in her presence. She made me feel safe.

An organization that assisted veterans helped me find a job as a research laboratory technician at the Markite Corporation, a Greenwich Village firm doing revolutionary work with a patented plastic that carried electrical current (plastics are normally insulators). I fabricated and tested potentiometers—voltage regulators—for the *Saturn 5* rocket. It was an interesting job until I got to know the process, which never varied and soon became rote. During lunch breaks, the technicians would gather in the lunchroom and play dominoes. I read instead.

One day in spring 1968, Dr. Scott Bonis, a research scientist whose lab I had been assigned to, sat down with me and asked what I was reading. I showed him my book on modern physics. He asked about my background. I told him I had been a physics major at Penn State and about my interest in the nature of time and Einstein's theory of relativity. I explained that I had decided to take a break from school.

"That's a mistake," he said. From then on Dr. Bonis insisted, almost daily, that I should return to school to finish my education and become a professional physicist. Before long, he decided to leave the company, and on his last day he took me to one side and made me promise that I would return to college sooner rather than later.

That evening when I returned to the Brooklyn apartment Dorothy and I shared I told her about my promise to my departing supervisor. Dorothy had found a job in the actuarial department of a Manhattan insurance company. She loved her work, and the vibrancy of New York City. But she said she had sensed my restlessness, and would support my decision to return to school. We talked it over and decided we would return to Pennsylvania come summer.

We were living in an upstairs one-bedroom apartment in Brooklyn. The building was a brownstone owned by a good-natured Jamaican woman who lived downstairs. On the evening of April 4 we heard her crying. Alarmed, Dorothy and I rushed to see what was wrong.

"They killed Dr. King," she sobbed. "Dr. King is gone."

The shock, anger, and frustration came pouring in. I vividly recalled the first time I heard Martin Luther King, Jr. It was summer, 1963—I was still in the air force—and the occasion was his soon-to-be-famous "I Have a Dream" speech in Washington, D.C. I immediately loved the deep resonance of his voice, and the way he used his words to touch and inspire. As I had so recently spent time in Mississippi, the sad truths of his message about what life was like in parts of America a century after emancipation hit home. "One hundred years later," he passionately intoned, "the life of the Negro is still sadly crippled by the manacles of segregation and the chains of discrimination." Somehow, he dispensed hope; with men like him in the forefront of the civil rights movement, it somehow seemed possible that our country might yet make good on the promise inherent in the Declaration of Independence that "all men are created equal." That hope was rewarded on the joyous occasion when the Civil Rights Act of 1964 was passed and signed into law. Not a first step—the Fourteenth Amendment to the Constitution should have been that—but a mighty big one.

And now the man who engendered such hope was gone.

That King's assassination was followed two months later by the murder of Robert F. Kennedy robbed me of whatever lingering hope I harbored as a black American that we were on the verge of racial equality. In those two tradegies, I felt that something fundamental had been lost in the soul of our country.

*   *   *

In the summer of 1968 I began classes at the main campus of Penn State University in the town of State College, forty-five miles northeast of Altoona in the geographic center of Pennsylvania.

At the foot of picturesque Mount Nittany, State College is a quiet town of some 38,000 permanent residents, most of whom are employed by Penn State. For its size, State College is a refreshingly cosmopolitan community with a downtown atmosphere reminiscent of Greenwich Village. I was delighted to be there, and back in school.

After working for a while in a department store, Dorothy landed a job in the university's undergraduate admissions office and eventually was promoted to executive secretary for the head of the department. She liked the work, and was delighted to be able to enroll at the university part-time at reduced tuition.

I returned to my studies with a renewed commitment. I was now able to take advanced physics courses and receive instruction in the physics of relativity and quantum mechanics taught by some of the finest professors at the school. Enthused by my studies, I made the dean's list for the first time that summer term.

I also made someone else's list, as I soon received a visit from the FBI.

The Watts Riots had happened while I was still in the air force, working the night shift and wrapped up in my own world of reading and studying. Of course, I was aware that the country seemed to be teetering on the edge of a race war. Status, power, and wealth were regularly denied many people of color; having experienced racism myself, I could empathize. And yet, as a law-abiding citizen and member of the military, I did not condone the street violence and loitering. It seemed to me there were a lot better ways to express frustration than burning down one's own neighborhood. Like Dr. King, I believed the majority of Negroes still supported nonviolent resistance to laws and practices that would keep us separate and unequal.

The FBI agents started off with saying they knew how "patriotic" I was due to my four years in the air force and my honorable discharge. They said there was a way I could still serve my country.

"How's that?" I asked, suspicious from the get-go.

"We would like you to infiltrate the black student movement on campus," one of the agents explained. "Make mental notes about what is going on during the meetings and who attends them. Report to us on a regular basis."

Shock must have registered on my face, because the agent immediately began soft-pedaling their request for me to be an FBI campus spy.

"We aren't really concerned about the movement per se," the agent went on. "We are basically looking out for the interests of all the students."

I was seething at their suggestion that I should spy on fellow students—black and white—and report their activities. I told the agents to forget it.

"Think about it for a few days and we'll talk again."

When they returned, I told the agents I hadn't changed my mind. The answer was still no. One of them then made this chilling statement: "Okay, you can continue on your career." The veiled threat was clear. They wanted me to believe that the FBI could, any time it wanted, derail my academic career.

This would-be recruiting incident made me even more interested in the black power movement. I took time out from my studies to read *The Fire Next Time,* written by James Baldwin two years before Watts. The book stoked my anger at the government's treatment of the black community, and I found myself more determined than ever to succeed in becoming a physicist.

One of the first courses I took on the main campus was modern physics, where I learned more of the details of Einstein's special theory of relativity. At last I began to fully understand why time slows down for a moving clock, a concept I first read about at age twelve in *The Universe and Dr. Einstein.*

The explanation given in the course was based on a device called a light clock, which consists of a vertical transparent tube with a mirror at opposite ends and a light beam that bounces back and forth between the mirrors. One tick of the clock occurs when the light beam goes from the bottom mirror to the top mirror and returns. If the clock is moved to the left or right, the light beam from the bottom mirror has to travel a longer distance to hit the upper mirror. As the clock continues to move, the light beam has to follow a longer path to return to the bottom mirror. Since the light beam is taking a longer time to get from the bottom mirror to the top mirror and back again, it takes longer for one tick of the moving clock to occur. So, time for the moving clock is longer than time for a clock that is standing still.

The proof that time slows down for a moving clock comes from experiments with cosmic rays. These rays are highly energetic elementary particles that constantly bombard the earth from all directions of space. Some of the particles in cosmic rays live only for an extremely short period before they disintegrate. One of the short-lived particles is called a muon. A muon is a heavier cousin of the electron, about 200 times more massive than an electron. As far as we know, electrons never disintegrate. On the other hand, a muon only lives for about one millionth of a second before it disintegrates.

Physicists were faced with a problem when it came to muons. Muons are created in the earth's upper atmosphere. Since the muons live for such a short period of time, they should disintegrate in the upper atmosphere and none of them should be seen at the earth's surface. But what physicists find is that we are showered at sea level with thousands of these muons. How do they live so long?

Without Einstein's special theory of relativity it is impossible to understand what is happening. The muons in the upper atmosphere are traveling close to the speed of light. Because of this high rate of speed, the internal clock of the muon slows down. Calculations using Einstein's special theory show that it is possible for a muon traveling close to the speed of light to live eight times longer than it would normally. Since the muon is living longer, it has time to reach the surface of the earth before it disintegrates, which explains why we are able to observe muons at sea level. Time for the rapidly moving muon has slowed down.

This "time dilation effect" also leads to a related prediction associated with the theory of special relativity. This prediction, called "the twin paradox," was a favorite of Einstein's and had caught my attention early in my readings on his work. The twin paradox goes like this: consider twenty-five-year-old twins, Fred and Jim, in the year 2035. Suppose that Jim is the more adventurous of the brothers and he takes a journey on a rocket that is going close to the speed of light. He journeys to a star that is twenty-five light years away—that is to say, it takes twenty-five years for light leaving the star to reach the earth. As for Fred, who has stayed behind on earth, he sees a time dilation effect for his twin. This means that as Jim's rocket approaches the speed of light everything that measures time on Jim's rocket slows down. This slowing down includes Jim's metabolism and heart rate. On the other hand, onboard the rocket nothing unusual is occurring for Jim. In fact, time passes for Jim at its normal rate, and he finds that the journey to the star takes only about five years. From Fred's point of view, it has taken Jim fifty years to make the round-trip to the distant star and back. Fred is now seventy-five years old when Jim arrives home in 2085. However, from Jim's standpoint the round-trip journey has taken only ten years, and he's only thirty-five years old. This means that Jim's rocket has acted as a time machine, in that it has taken Jim only ten years to travel fifty years into the earth's future.

Scientists have long known with certainty that this type of time travel to the future will occur, because we have already seen it in connection with real experiments involving subatomic particles like the muon. Someday when technological advances have led to propulsion systems capable of attaining velocities near the speed of light, effects such as the twin paradox will be a normal occurrence.

We will by then, I hope, have taken into account the sociological implications associated with an astronaut leaving home and traveling near the speed of light and returning to find that he is younger than his grandchildren.

This type of time travel associated with the time dilation effect in special relativity is a one-way trip. For example, when Jim arrives on the earth in 2085 he can never get back to 2035. So, Einstein's special relativity allows time travel to the earth's future but not into our past—the direction in which, for my own very personal reasons, I was interested in going.

I also learned in the modern physics course why nothing in the universe can exceed the speed of light. As the speed of an object increases, its mass increases. The reason for this can be found in the most famous equation of Einstein's special theory of relativity: $E = mc^2$. Where E is the energy, m is the mass, and c^2 is the speed of light multiplied by itself. This equation says that if you apply energy to speed up an object, some of that energy goes into becoming a part of the mass of the object. As the object gets more massive it becomes harder to speed it up. Eventually, it would take more energy than there is in the entire universe to get the object to go at the speed of light, which means we can never speed up an object to reach the speed of light. We can get *close* to the speed of light, but we can never reach or exceed it. The speed of light is the universe's maximum speed limit.

The course discussed Einstein's general theory of relativity as well, which is also considered Einstein's theory of gravity. I learned that Einstein had to invent the general theory of relativity because of a problem with Newton's theory of gravity developed in the seventeenth century. In Newton's theory, the gravitational force of attraction between two objects, like the sun and the earth, become stronger the closer to one another you bring the two objects, and weaker the farther you move them away from each other. But something strange happens when you compare the gravitational force with the speed of light.

The earth is 93,000,000 miles from the sun. It takes light, traveling at 186,000 miles per second, a full eight minutes to get from the sun to the earth. Should some cosmic catastrophe occur that destroys the sun, here on the earth we would still see the sun for eight minutes before it disappeared. However, according to

Newton's theory, the gravitational force shuts down immediately and the earth would no longer feel the attraction from the gravity of the sun. This would mean that even though we could still see the sun in the sky for eight minutes, the gravitational force would go away immediately. Sans gravity, we could find ourselves hurtling into space even though we could still see the sun. Einstein said this cannot happen because it implies that the gravitational force can travel faster than the speed of light. The only way that Einstein could solve this problem was to develop a whole new theory of gravity.

Einstein's new theory of gravity can be understood by a simple example. Consider a rubber sheet stretched on a wooden frame—something like a trampoline—with the rubber sheet representing empty interstellar space. If a bowling ball is placed on the rubber sheet it will curve the sheet. If we then place a little marble on the rubber sheet at some distance away from the bowling ball, the marble will roll along the curved rubber sheet until hitting the bowling ball. Now, imagine that the rubber sheet is transparent and we can see only the marble and the bowling ball. The situation will now look to us as though the bowling ball is directly attracting the marble due to some force of attraction between them. However, this is an illusion. What's really happening is that the bowling ball is curving the rubber sheet and the marble is moving along that curvature toward the bowling ball.

Einstein said that this is the sort of thing that's happening in empty space. In other words, a massive object like the sun is bending the empty space around it, and the earth is moving along the curved empty space created by the sun. If the earth didn't have its sideways motion it would plunge into the sun just the way the marble plunges toward the bowling ball. But because of it sideways motion, the earth is moving constantly around the sun like one of those roller-derby skaters Grandpap liked to watch go round and round the rink.

Einstein's theory of general relativity states that the gravitational force of attraction is really the bending of empty space by a massive object. With this new view of gravity as the bending of space, Einstein was able to reconcile the speed of light with the gravitational force. In his theory of gravity, the destruction of the sun by some cosmic catastrophe would change the bending of space, but it would take the change in the bending of space eight minutes to reach the earth. As long as we can see the sun, we would feel the gravitational force due to the sun. Gravity is the bending of space, and that bending cannot travel faster than the speed of light.[10]

One of the most important experimental tests of Einstein's general theory of relativity tested a prediction that Einstein made about what happens to a ray of light when it comes close to the gravitational field of the sun. Einstein predicted the behavior of a star that is located behind the sun. If the star is directly behind the sun, we can't see it because the light from the star is blocked by the sun. However, Einstein said that, because the sun is bending space, a ray of light from the star passing near enough to the sun will get deflected by the curved space around the sun. We would be able to see the deflected ray of light. It would appear as though the star is not behind the sun but is located at the edge of the sun.

Einstein's prediction of the deflection of starlight by the sun was verified in 1919 in one of the most famous observational experiments in the history of physics. The observation was performed by Englishman Arthur Eddington (1882–1944), one of the most prominent astrophysicists of his time and among the first scientists to appreciate the importance of Einstein's relativity theories. In the night sky, Eddington located stars that would normally be blocked by the position of the sun. Then, during a total eclipse of the sun, he measured the positions of the same stars near the edge of the sun. Eddington found that the apparent positions of the stars had shifted from their actual positions by just the amount predicted by Einstein. At a meeting of the Royal Society in England in 1919, Eddington announced that Einstein had correctly predicted the deflection of starlight due to the gravitational field of the sun.

The news of Eddington's successful observation of Einstein's prediction caused an international sensation and was the beginning of Einstein's worldwide popular celebrity. This was due in part not only to Einstein's strange new theory of curved space, but also to the hope that people found in the scientific collaboration of two nations—England and Germany—that had so recently been enemies in one of the most horrific wars the world had ever seen.

The Penn State modern physics course had not gone into the equations of general relativity, which had so baffled me when I first saw them in the paperback edition of *Principle of Relativity*. To understand the equations, one needed to follow in Einstein's footsteps and learn a whole new mathematical technique called tensor calculus.

After completing his special theory of relativity, Einstein turned to the problem of gravity, but soon realized that the mathematical tools that he had been using were not adequate for the task. The mathematics he had used in special relativity was based on ordinary calculus, which can be thought of as the mathematics of motion in flat space. Newton had developed calculus as a means of calculating the motion of planets. In using ordinary calculus, it is assumed that space is flat and unchanging.[11]

In contrast, tensor calculus is the mathematics of motion in curved space. It works and looks the same on any kind of curved space. An integral segment of tensor calculus is its use of vectors, defined as directed line segments. A vector is represented by drawing an arrow as a straight line. The direction of the arrow is the direction of the vector, rather like a straight arrow drawn on a map between two cities. A tensor is a generalization of a vector, taking into account the possibility of representing a quantity that is pointing in more than one direction at the same time.

An example of a tensor would be the stresses on a tennis ball that is being squeezed by a mechanical vice. The different parts of the tennis ball are experiencing different stresses. A single quantity like a vector cannot represent the stresses on the ball coming from multiple directions; only a tensor can represent such forces.

In Einstein's general theory of relativity, matter causes stresses in space, which result in a curvature of space, which we commonly call gravity. For this reason, Einstein needed to use tensors for this new theory of gravity. (While tensors were explored before Einstein, the acceptance and success of his general theory of relativity led to their widespread exploration and use by mathematicians and physicists.)

To learn tensor calculus, Einstein sought the aid of his long-time friend, Marcel Grossman, who was a professor of mathematics in Zürich, Switzerland. It took Einstein ten years, from 1905 to 1915, to learn tensor calculus and complete his revolutionary theory of general relativity. In tensor calculus, Einstein's famous equation of the gravitational field of general relativity has the following form:

$$R_{\mu\nu} - \tfrac{1}{2} g_{\mu\nu} R = k T_{\mu\nu}$$

The left-hand side of this equation shows the curvature of space; the right-hand side represents the stresses of matter. The equal sign in the equation means that the stresses of matter cause the curvature of space.

Fortunately, since Einstein's day, courses in tensor calculus have been developed that have taught future generations of mathematicians and physicists. Given my interest in pursuing general relativity, I decided that I should take the tensor calculus course that was offered by the mathematics department at Penn State. Already I had been taking so many courses in mathematics, along with my physics courses, that some members of the physics department thought I was a math major. I found tensor calculus an easy language to learn, which put me on track to next tackle the technical aspects of general relativity.

It was in the modern physics course that I finally learned the meaning of Schrodinger's equation, which had so moved me with its symmetrical beauty when I first came across it in a correspondence course while I was in the air force. My determination to one day understand Schrodinger's equation had surely led me in the direction of studying physics, and even to my eventual career path.

In Schrodinger's equation, there is the Greek symbol psi: ψ. This symbol is called the wave function and does not describe where an electron actually is but only where it might possibly be. At the core of quantum mechanics is the uncertainty in the behavior of the electron represented by the wave function. Under any given circumstance, we cannot predict exactly what the location of the electron will be; we can only predict the probability of it being at that location. This interpretation of Schrodinger's wave function was first introduced in 1926 by the German physicist Max Born (1882–1970). When the wave function is multiplied times itself in a special way that looks like $\psi * \psi$, we can then determine the probability of the electron's location. The larger the number that we get from $\psi * \psi$ at a particular location, the more likely the electron will be found at that specific location.

As I thought about this feature of quantum mechanics, it seemed strange because it implied that the electron has some probability of being *anywhere*. I soon discovered that this sense of strangeness was shared by none other than Einstein. While reading Ronald Clark's comprehensive biography *Einstein: The Life and Times,* just a few years after I took the modern physics course, I learned that Einstein had been strongly opposed to the probabilistic interpretation of quantum mechanics. Einstein fundamentally disagreed with the developers of quantum mechanics, such as Niels Bohr, Werner Heisenberg, and Max Born. For Einstein, the world was deterministic—that is, in any given situation one can exactly determine the behavior of an individual electron.

According to Einstein, statistics can and should be applied to situations in which one is dealing with a large number of particles. For example, only the average behavior of the millions of molecules in a gas can be determined. In fact, the average energy of the motion of gas molecules is what we call temperature. Einstein was a master of using statistics to determine the behavior of physical systems. In his famous explanation known as Brownian motion— named after English botanist Robert Brown (1773–1858), who in 1827 observed that pollen grains suspended in water jiggled about under the lens of his microscope in a zigzag path—Einstein showed that the seemingly random motion of the pollen grain suspended in a liquid was due to the grain being bombarded by millions of water molecules. Prior to Einstein's definitive work, physicists had been skeptical about the existence of molecules, the smallest particle of a pure chemical substance that still retains its chemical composition and properties.[12]

Although Einstein's work on the photoelectric effect—which showed how electrons are ejected from metal by being bombarded by particles of light called photons—was one of his early great contributions to quantum theory, when it came to quantum mechanics he was a reluctant revolutionary. He could not believe that statistics were needed to explain the action of an individual electron. Einstein's famous quote regarding the myriad probability factors involved in quantum mechanics states: "God does not play dice with the world."

With my academic career back on track, Dorothy and I decided to marry in 1969. Since our return to Pennsylvania, she had finalized her divorce. Our civil wedding ceremony was not attended by anyone from either of our families. Dorothy's father, Ken Fry, who was displeased about her marrying a black man, announced that he did not wish to see her again. Fortunately, he eventually relented, and he and I developed a good relationship. Ken had wanted to become an engineer but after returning from World War II had gone to work running a boiler at a paper mill to support his young family and was unable to return to school. He read engineering journals, and enjoyed discussing technical theories with me. It was a valuable lesson for me in how prejudice can be overcome by getting past racial generalizations and connecting as individuals. Ken and I ended up respecting and even liking one another. Still, I never felt comfortable telling him about my interest in time travel. In retrospect, I imagine that, as a thwarted, would-be engineer, he would have been enthralled with the technical challenges of my dream project.

Dorothy and I joined the University Baptist Church, and Dorothy became a member of the choir. The church became a large part of our social life in State College. None of us there had much money, and the parties at the homes of friends were always casual and simple. Once a month we would get together with friends and go out for pizza and beer, and that was something special. As far as how other people viewed Dorothy and me as an interracial couple, we didn't notice anything negative. We had our friends, families, and each other.

To help make ends meet, I took a part-time night job as a computer programmer for a trucking company, which eventually offered to pay for my schooling if I would switch my major to computers and work for them at least one year after graduation. It was a generous offer, but I could not fathom abandoning my studies in physics. I still received the GI Bill, which covered my tuition and books, and I occasionally taught as a substitute at the local high school, an experience that made me realize that one day I wanted to teach physics in college in addition to conducting my own original research.

I graduated with my bachelor's of science degree in physics in the winter of 1969, and went straight into a master's program in physics, which I completed in December, 1970. As I was finishing my master's work I began to look around at the Ph.D. programs of other institutions. (On average, a Ph.D. takes four years of full-time work, although, depending on the student's desire, aptitude and ability to work independently, it can take much longer.) It was then that the associate dean of Penn State's College of Science asked to speak to me. He said that because of my academic record during the time I had been on the main campus, he was nominating me for a National Science Foundation Traineeship. If I received the NSF appointment, it would mean that I would receive financial support during my Ph.D. program and be able to devote all my time to research without having to worry about the usual duties associated with a paid teaching assistantship or having to find other work to supplement my income. However, the traineeship was nontransferable; by accepting, I would have to remain at Penn State, which I ended up doing not only for financial reasons but also because the university and the physics department had treated me well.

I knew just the faculty member I wanted to work with on my Ph.D. thesis: the brilliant, iconoclastic, and charismatic Professor Gordon Fleming. Fleming had a top-flight reputation as both a teacher and a researcher. He was a free spirit with movie-star good looks and a deep baritone voice, and he was popular with students and staff alike. At one point, Fleming had the largest number of graduate students of any of the physics faculty doing theoretical research. He enjoyed cigars and riding his motorcycle. A true renaissance man, he was as much at home in the philosophy department as he was among physicists. I had heard of Gordon Fleming long before I met him. It was rumored that his nickname Flash Gordon came about because of the speed that he could write equations on the blackboard.

In the course of working on my master's, I had given a lecture about the Einstein-Infeld-Hoffman (EIH) problem at a physics seminar held at Penn State. The EIH problem is a technical one that goes to the roots of the general relativity theory. General relativity normally required that one set of equations be used to determine the geometry of curved space created by a massive body like the sun, and a separate set of equations to determine the motion of a body like the earth in the curved space created by the sun. In 1938, Einstein and his collaborators, Leopold Infeld and Banesh Hoffman, were able to show that only a single set of equations was required to determine the geometry of curved space created by the sun. This single set of equations can be used to directly determine how a body like the earth moves around in the geometry created by the sun, replacing the need for two separate sets of equations— one set for the geometry and another set for the motion.

By this time in my education, I had discovered that I could remember and reproduce long mathematical derivations. The type of memory I have seems more temporal than spatial; that is, I remember equations and derivations like a musical theme rather than visualizing a whole equation as if it were a photograph. So, for my seminar presentation, I decided to derive and write all the equations for the EIH problem on the blackboard from memory. I knew that Professor Fleming would be in the audience, and I must admit I did so partly to impress him with my quickness and agility with equations. During the question and answer period, a faculty member asked me if I had given any thought to whom I would like to work with on my Ph.D. thesis. "Dr. Fleming," I said without hesitation. I noticed that Fleming smiled rather proudly.

I was also interested in working with Fleming because of a paper he had written entitled "Timelike Reflection: The Coupling between Time Reversal and the Poincaré Group." When I saw the term "time reversal" in the title of the paper, I wanted to know more about Fleming's research.

The earlier work that he had done involved looking at what happens to the motion of a particle when you reverse the sign of time from *t* to –*t*, which amounts to putting the motion of the particle in reverse. That is to say, if the particle is moving from the left side of a room to the right side, changing the sign for time from *t* to -*t* would cause the particle to move back in the opposite direction. In this case, time reversal amounts to motion reversal. What Fleming was trying to achieve was a way to look at this time reversal from the standpoint of Einstein's special theory of relativity. In Einstein's special theory one tries to write equations in a way that everyone can agree on no matter how fast particles are moving with respect to each other. To do this, physicists found that they had to make use of a four-dimensional point of view that was developed by the German mathematician Herman Minkowski (1864–1909).

In 1908, three years after Einstein developed the special theory of relativity, Minkowski, Einstein's former math professor, realized that if you combined space and time into a new fourth dimensional entity called spacetime, the equations of physics could be rewritten in a simple way. "From henceforth," Minkowski wrote, "space by itself, and time by itself, have vanished into the merest shadows and only a kind of blend of the two exists in its own right." This new way of writing the equations of physics looked the same no matter how fast numerous bodies were moving with respect to each other, all of which takes place in the flat spacetime of special relativity. This perfectly clear, four-dimensional way of writing equations is called "manifest covariance."

The notion of manifest covariance was my first official introduction to the idea of elegance in physics. I was to learn that, for a trained physicist, elegance meant simplicity in the form of the equations; at the same time the physical content of the equations are clearly conveyed, there is also a certain artistic beauty in the form of the equations themselves.

Fleming, who felt that elegance through manifest covariance was something that one should try to achieve as well as correctness in physical theory, had successfully found a way to write the mathematical operation of time reversal, going from t to $-t$, in a way that everyone could agree on no matter what constant speed bodies were moving relative to each other. In other words, he had found a clear—and yes, elegant—way of describing time/motion reversal in Einstein's special theory of relativity. Fleming did this by using Minkowski's four-dimensional way of looking at things.

I was more interested than ever in Einstein's general theory of relativity, and wanted to know how gravity might affect the operation of time reversal. A short time after my seminar presentation, I approached Fleming and asked if I could do a Ph.D. thesis with him on the topic of looking at the effect of gravity on time reversal. He said he thought the topic would make for a good thesis, however, he cautioned that trying to tackle the general problem of gravity and time reversal was too broad. He suggested that we narrow the problem down to a particular curved space that was of interest to cosmologists. The space we decided on was de Sitter space. The curved space of de Sitter had a long and interesting history in cosmology.

In 1917, Einstein published a paper, "Cosmological Considerations on the General Theory of Relativity," in which he laid out his first attempt to apply his general theory of relativity to the universe as a whole. He unveiled a theory that the universe we see is static. To ensure that his equations would lead to that underlying assumption, Einstein went on to introduce a term into his equations that kept the universe unchanging. This concept he called the "cosmological constant," a genie he would regret letting out of the bottle following the groundbreaking work of American astronomer Edwin Hubble (1898–1953).

Hubble, who earned a degree in mathematics and astronomy at the University of Chicago, observed in the early 1920s at the Mount Wilson Observatory in Pasadena, California, that the galaxies surrounding our Milky Way are moving away from us in such a way that the greater the distance from us, the faster they are moving.

This motion occurs, Hubble recognized, in every direction around our galaxy. His conclusion was that the universe is constantly expanding—a bold contradiction to Einstein's static-universe theory. Einstein soon confessed to what he called his "greatest blunder." In an effort to agree with Hubble's astronomical findings, Einstein quickly dropped his theory of the cosmological constant, showing, if nothing else, that great minds remain nimble.

Dutch astronomer Wilhelm de Sitter eventually solved the gravitational field equations of Einstein's general theory of relativity. De Sitter postulated that even with the cosmological constant the universe could still be shown to be expanding. It turned out Einstein was wrong not only about the universe being static, but also about his concern that the cosmological constant was a problem for his general theory of relativity. The solution of Einstein's gravitational field equations leading to an expanding universe with the cosmological constant is now known as the "de Sitter solution," and the universe it represents is either "de Sitter space" or the "de Sitter universe."

I soon learned that there is a simple way of imagining the expansion of de Sitter space. Take a black balloon and paint white spots on it—the balloon represents empty space and the white spots represent the galaxies in the universe. As one blows up the balloon, the white spots start moving away from each other. No matter what spot the eye focuses on; all we see are other spots moving away from that spot. The expansion of the balloon represents the expansion of the universe. Like a balloon, the universe is all surface. (Also like a balloon, the universe has a finite size. There is an equation for the size of the universe; its radius is believed to be some fifteen billion light years.) The motion of the balloon's spots away from each other represents the motion of the galaxies away from each other. And the spots are moving away because they are on the surface of the balloon. In just the same way as the spots on the balloon, the galaxies are moving away from each other because they are embedded in space. It is space that is expanding (like the balloon), and the galaxies are along for the ride (like the stars on the balloon). This is the expanding universe that Hubble first observed with his telescope.

My thesis was to find a way to describe time reversal in de Sitter space. It turned out that stating the thesis was far simpler than solving the problem. In fact, I spent months with approaches that led to dead ends. After nearly a year and a half, I began to despair of ever finding the solution to the problem. Then, one afternoon I came back to my apartment exhausted and lay down on the couch. As I dropped off to sleep, I had this dreamlike vision of mathematical symbols putting themselves together in different ways until it seemed that the symbols for four dimensions combined with an extra symbol to make a fifth dimension. I awoke with a start and realized that the four dimensions of flat Minkowski space of special relativity were not enough. What I had to do was to go to five dimensions. I had to use the four dimensions of spacetime in an extra dimension that represented the curvature of de Sitter space in a fictitious hyperspace. I knew at once that this five-dimensional point of view was correct.

Excitedly, I wrote up my results and took them to Fleming. He said this approach looked "promising," but he wanted to study it. I said that if he felt my results were correct, which would be a major breakthrough to completing my thesis, I would very much like to have them appear in the *Journal of Mathematical Physics,* a respected and refereed professional journal.

Publication was not a requirement of the Penn State physics department for Ph.D. students, and Fleming cautioned me that my paper could be rejected. He asked how I would feel if that happened. Not good, obviously. Nevertheless, I told him I was willing to take that chance.

I had a couple of reasons for this request. The first was due to an incident that happened while Dorothy and I were living in

Brooklyn. One afternoon we went to the Brooklyn Public Library, and as I browsed through the physics journals I picked up an issue of the *Journal of Mathematical Physics,* published in a beautiful, red velvet-like cover. At that point, I couldn't understand any of the technical papers in the journal, but I told Dorothy that one day I would understand what these scientific papers meant and have my own paper published in this handsome-looking journal.

The other reason was more complex. I wanted proof outside of my department that I was a professional physicist by having the paper based on my thesis published in a refereed journal, meaning that a leading scientist in the field would decide its originality and worthiness. That way there could be no question that my thesis was original and worthy of a Ph.D.

I didn't hear anything from Fleming for a number of weeks. Then, at the physics department picnic, Fleming sauntered over to me and as he gnawed on a barbecued chicken leg said that everything looked good and he had gone ahead and submitted the paper for publication as I had requested. Then he walked away as casually as if on a springtime stroll through Central Park. Speechless, I turned to Dorothy, who rewarded me with her gorgeous smile, accompanied by a flow of happy tears. We hugged like teenagers.

My thrill soon passed into a state of anxiety. Waiting for the report from the referee made the summer of 1972 one of the longest of my life. Finally, word came that the referee considered the paper original and interesting, and with a few minor changes it was recommended for publication. My first academic paper, "Position Operators in a (3 + 1) de Sitter Space," was published in the *Journal of Mathematical Physics* in January, 1973.

I brought home a copy of the published paper and autographed it to Dorothy, thanking her for her love and support. "This is just like winning a gold medal at the Olympics," I gushed. A few days later, Dorothy presented me with a shiny gold medallion at our own private awards ceremony.

As I would ultimately discover about myself, my personal happiness is greatly tied to whatever is happening in my work. In graduate school, with my thesis research going well, I was on top of the world. For our summer vacation, Dorothy and I took a road trip to Columbus, Ohio, to visit a science museum. We both loved any reason to travel, and I always took physics books with me. I recall feeling such a sense of joy as I worked calculations while Dorothy drove across long stretches of interstate highway. I always made a point of explaining my latest work to her in elementary ways with analogies that someone without a background in advanced science and mathematics could grasp, and she seemed to appreciate the explanations. In a real sense, I think this interaction between us helped prepare me for my future role as a classroom teacher. On our drives, we listened to Peter, Paul, and Mary tapes and sometimes sang along. These were very happy times.

I spent the next academic year completing my thesis. The second paper to come out of my thesis showed how to write the operation of time reversal in a way that everyone in the expanding de Sitter universe could agree on. The paper was entitled, "Coupling between Time Reversal and the Space-Time Symmetries of the de Sitter Universe." This paper was my generalization of Fleming's work applied to time reversal in a curved space, which was eventually published in another professional journal, *Physical Review.*

After two and a half years of study, I received my doctorate in physics from Penn State in 1973. At the time, I was one of only seventy-nine African-American Ph.D.'s in physics—out of approximately 20,000 Ph.D. physicists in the United States. (As of 2006, there are approximately 250 African-American Ph.D. physicists in the United States out of a total of 25,000. No black scientist has yet won a Nobel Prize in any field.)

My career in physics, and the next stage of my pursuit of time travel, was about to start in earnest.

Seven

My Introduction to Lasers

⧗

Now that I had my Ph.D. it was time to find a job.

Even before I finished my thesis I had started searching for a position in physics. I hoped to continue my research in general relativity and explore the new avenues that might open up regarding time travel to the past. In order to accomplish this, I needed to find an academic position at a college or university that would allow me to conduct research as well as teach. Unfortunately, the country was in a recession, and there was decreased funding for the sciences.

I searched the classified section of the American Physical Society's trade journal, *Physics Today*, but the pickings were slim. Previously, I had seen issues with several pages of physics openings, but now with the economy in trouble there was only one side of one page. It was not encouraging. Nevertheless, I went to the Penn State career center to look up addresses and applied to nearly a hundred colleges and about a dozen corporations. I never heard from any of the colleges, but to my surprise I received six responses from industry requesting an interview.

Becoming an industrial scientist was not what I had ever envisioned. However, the advantage to being a trained physicist is that one develops analytical problem-solving skills that can be applied to other areas of science and even engineering, as physics is at the root of all of the engineering disciplines. With this in mind, I felt equipped to apply my knowledge in an industrial environment, although my hope was to continue to search for a way into academia while working in industry. I took heart in remembering that even Einstein had been unable to find a teaching job upon graduating from college and had been required to take a position in the patent office to pay the bills.

My first interview was at General Electric in Schenectady, New York. It was the middle of the winter and the snow was knee-deep in some places. The people at the G.E. laboratory were cordial, but I wasn't thrilled about winters here. My next interview was at the G.E. think tank, Tempo, in Santa Barbara, California. I liked the area and the people I met, and the offices were located just a couple blocks from the beautiful Pacific Ocean, but I couldn't get a read on how the interview had gone. When I returned home a bit discouraged, I learned that United Technologies (at the time, United Aircraft Research Laboratories) in East Hartford, Connecticut, wanted to interview me.

United Technologies consists of several divisions, the largest being Pratt & Whitney Aircraft, which produces engines for commercial and military aircraft. The Research Center is a separate division (modeled after Bell Laboratories) that is engaged in scientific research aimed at industrial application. My interview was with Dr. Gerald Peterson, principal research scientist in charge of the Theoretical Physics Group. I liked Jerry Peterson right away. A tall, lean man with an open engaging manner and a quick, eclectic mind, he had received his Ph.D. in theoretical solid-state physics from Cornell University. The members of the group he headed were first-rate scientific problem solvers who were contracted out as needed to the other divisions of United Technologies.

I was a little self-conscious and probably looked like a genuine nerd. I even had a strip of white tape holding my horn-rimmed glasses together until I could afford new ones. Peterson put me at ease and asked insightful questions about my thesis research and scientific interests. He then asked if I felt I could do industrial scientific research.

"My thesis adviser said that a good theoretical physicist could do anything," I replied, not realizing how arrogant that may have sounded.

We had lunch in the cafeteria, and Jerry introduced me to some of the other members of his scientific group. Overall, I felt that the interview went well, and I told Peterson that I liked the environment of the Research Center. I also mentioned that in a couple of days I would be flying back to California for an interview at TRW in Redondo Beach.

I returned to Pennsylvania that evening. The phone rang before I had finished unpacking my bag. It was Jerry Peterson calling to offer me a research scientist position in his Theoretical Physics Group. Years later, Jerry would tell me he decided to hire me when I described how I had made the breakthrough regarding my thesis—the afternoon I had gone home and taken a nap on the couch and dreamed of a solution. He figured that anyone who could work out physics solutions in their dreams was someone he wanted on staff.

Dorothy and I both liked the idea of Connecticut and its closeness to New York City. I accepted at once, and we packed up our belongings and moved in the summer of 1973. We settled into an apartment complex recommended to us by United Technologies, where other new employees and their families were living. Dorothy stayed home the rest of the summer, reading, working on a tan, and watching the neighbors. She shared with me one funny scene that she said took place at precisely 5:00 PM every day at the apartment patio directly across from us. The neighbor came out, opened the barbecue, took off the grill and placed it on the ground, then lit the charcoal. She then went inside and her dog went over and licked the grill clean. When the fire was ready, the lady would come out, put the grill back on the barbecue, and place the evening meal atop it. When Dorothy showed me the scene unfold exactly as she had described it, my comment was a quiet, "Yuk!" It took us a while before we were ready to buy a barbecue.

My first assignment was a contract with Pratt & Whitney. At the time, Pratt & Whitney was investigating the possibility of using high-powered lasers to drill holes in the turbine blades of their jet aircraft engines. They had been performing experiments and needed theoretical support. As the theoretical results of my Ph.D. thesis had been very remote from experimentation, this was the first time I had been in a situation where the theory I would be developing could be immediately tested by experiment. That prospect was a bit daunting.

My assignment was to develop a mathematical model for the laser drilling of metals and alloys of interest to Pratt & Whitney. At this point, I had practically no knowledge of lasers. However, as a Ph.D. researcher I had been trained to do creative problem solving, and Jerry said I would be given the time to learn the necessary background. So I spent weeks in the company library reading about lasers, which at the time was a relatively new light source.

The world's first operational laser was built in 1960 at the Hughes Research Laboratories in Malibu, California. Lasers, however, would never have been possible without the development of quantum theory, so credit must first go to the German theoretical physicist, Max Planck (1858–1947), who at the dawn of the twentieth century developed the quantum theory, for which he received the 1918 Nobel Prize in Physics.

In 1900, Planck introduced a radical explanation of how energy is exchanged from the heat radiation in an oven to the walls of the oven. He said that the energy does not flow continuously from the heat radiation into the walls but is given to the walls only in definite amounts, or quanta. Prior to Planck, physicists thought that the exchange of energy from heat radiation was like pouring water into a glass, although this did not agree with what had been observed experimentally. Planck proposed that the energy exchange was more like putting coins in a slot machine, with each coin being a separate quantum. Planck's theoretical model was soon verified through experimentation. Because energy was exchanged in quanta amounts, the new theory was called quantum theory. It was the beginning of a revolution that was to have an impact on our entire understanding of matter and energy, and lead to a technological revolution that is still developing today.

In Einstein's miracle year of 1905, when he published his special theory of relativity, he also made a major contribution to quantum theory. Einstein suggested that not only is the energy exchanged in definite quanta but the radiation itself inside the oven is also made up of individual particles of light called photons. Einstein used his theory to explain the phenomenon known as the photoelectric effect. The German experimental physicist Philipp Lenard (1862–1947) observed that shining light on a metal surface caused electrons to be ejected from the surface of the metal. Einstein was able to explain this result by saying that the light shining on the metal surface consisted of photons bombarding the surface. It was these photons that caused electrons to be ejected from the metal. It is for this work, and not relativity, that Einstein won the 1921 Nobel Prize in Physics. Today, the photoelectric effect is important for everything from automatic door openers to television camera tubes.[13]

Einstein's work led to the next important development in quantum theory and provided another piece of the laser story, when, in 1913, Danish physicist Niels Bohr used Planck's quantum theory and Einstein's theory of photons to explain the behavior of the hydrogen atom. As it happens, prior to quantum theory physicists could not understand why the hydrogen atom didn't collapse on itself. A hydrogen atom consists of a positive proton orbited by a negative electron. As the electron revolves around the proton it should continuously lose energy and eventually collapse into the proton. Since this is not what happens, something else had to be going on. Bohr suggested that the key to the problem was quantum theory. In quantum theory, the energy cannot be lost continuously, as energy is not continuous. This means that the electron has to stay in a fixed orbit, called its energy level, because it cannot lose energy continuously.

There's one final ingredient in the theory behind the laser. Once again this was provided by Einstein, who in 1916 published a paper entitled "Emission and Absorption of Radiation According to the Quantum Theory," in which he combined the ideas of Planck, Bohr, and his own earlier work. Using this combination of ideas, Einstein investigated the three fundamental processes by which atoms emit and absorb light. The first process is called induced absorption. In this process, a photon coming into the atom is absorbed by an electron in a low energy level. The incoming photon induces the electron to jump to a higher energy level. The second process is called spontaneous emission. Once the electron is in the higher energy level it can spontaneously jump to a lower energy level by emitting a photon. This spontaneous emission process occurs randomly. The final process is called stimulated emission; if an incoming photon finds that the electron is already at a higher energy level, then the photon can kick the electron out of that level. In the process, the incoming photon doesn't get absorbed and keeps moving on. However, when the electron falls to the lower energy level it emits its own photon. So the net result is that a single photon coming into an atom can result in two photons coming out. One of the outgoing photons is the original photon and the other outgoing photon is the stimulated photon. This leads to an amplification of the light coming out of the atom, another key ingredient to understanding the laser.

The word *laser* itself reveals how they work. Laser is an acronym that stands for Light Amplification by the Stimulated Emission of Radiation. The first operational laser was designed in 1960 by an American physicist and junior employee at Hughes Aircraft, Theodore Maiman. He used a synthetic ruby crystal shaped into a cylinder. The trick was to get all of the electrons in the atoms of the crystal into a higher energy level at the same time, a process called population inversion. This is done by wrapping the ruby crystal with a flash tube. Light from the flash tube puts all of the electrons into a higher energy level. Two mirrors are placed at the opposite ends of the ruby crystal. One of the mirrors is completely reflecting and the other mirror is partially reflecting. A spontaneously emitted photon from one of the atoms gets things moving. That photon will stimulate the emission of another photon in a different atom. Then those two photons will hit two more atoms and stimulate those atoms to emit photons. Now there are four photons which stimulate four other atoms to emit, and so on. This will result in a chain reaction of photons that bounce back and forth between the two mirrors of the ruby crystal. Because each photon in the laser is in step with all the other photons that were stimulated by them, the beam they form is said to be coherent. There is an essential difference between the coherent photons of a laser source and photons from an ordinary light source such as a standard lightbulb. The photons from a lightbulb tend to clump together or bunch up. By contrast, the photons of a laser source are distributed like drops in a steady rainfall. The coherent state of a laser source was first described in detail by Roy Glauber.[14] It is the coherence of the laser output that allows a very narrow beam to be formed. Eventually the photons will stream through the partially reflecting mirror, resulting in an intense, incredibly narrow, pure laser beam of light. When a ruby crystal is used, the color of the laser light is red.

A distinctive characteristic of the laser is that the coherent narrow light beam can result in an intensity of millions of watts per square centimeter. At these levels of intensity, a laser beam on a metal surface will cause the temperature of the metal to rise to a point of vaporization. This would give the laser the ability, at least theoretically, to drill holes through solid metal blocks.

With my new understanding of how a laser worked and what it could do, I was now ready to develop a mathematical model that would predict how effective a laser beam would be in drilling holes. It turned out that to develop a realistic model it was necessary to calculate the energy transfer from the laser beam to the metal's surface, and then calculate heat conduction in the metal while the surface is moving. This is called a moving-boundary problem, and it can be notoriously difficult to solve. In fact, Pratt & Whitney had a team of engineers using elaborate computer codes to try to solve the problem. I planned to consult with them, but I found out rather quickly that information exchange within the company was not encouraged. This was quite different from the academic environment I had been accustomed to. In a university research setting generous information sharing was the norm, but I found that was not the case in private industry. In fact, Pratt & Whitney provided me with an early version of a desktop computer so that I would not have to use the mainframe computer and inadvertently let other people know what I was doing.

While browsing through the Research Center's library, I had hit upon a technique that allowed me to develop a mathematical model that would predict what would happen when a laser beam is drilling through metal. Using the model I developed, I was able to calculate accurately the exact depth of the hole that would be produced in a particular metal or alloy for any given laser intensity. The results of my calculations were soon observed experimentally during laser hole-drilling in the Research Center laboratory. At that point, my job was done.

While everyone seemed happy with my work on my first assignment, I had not given up hope of finding a way into academia. In the meantime, I continued my own studies. In the evenings, I would come home, have dinner with Dorothy, who after her summer of relaxation had found a job with a Hartford insurance company, and spend a few hours studying general relativity and the new theories that were being developed to understand the basic forces of nature.

Physicists believe that there are four basic forces in the universe. The strongest of these forces is called the strong nuclear force. This is the force that holds protons together with neutrons in the nucleus of atoms. The second force, in order of strength, is the electromagnetic force. Besides gravity, the electromagnetic force is the one that we most commonly come in contact with in our daily lives. The magnetic field of the earth, which determines the direction of a compass needle, is part of the electromagnetic force. This force also leads to the electrical attraction between protons and orbiting electrons in an atom. Radio and television signals are the result of changing electromagnetic fields. The next force is the so-called weak nuclear force. This is the force behind radioactive decay. Surprisingly, the weakest of the four forces is the gravitational force, which holds the planets in orbit around the sun and keeps us attached to the surface of the earth. It seems odd that gravity is so weak, but it depends entirely on the gravitational attraction between massive objects. The electrical attraction between the proton and electron is far greater.

Einstein spent the last years of his life trying to combine the electromagnetic force with gravitational force in what he called a "unified field theory." Later, I learned Einstein's striving for a unification of the electromagnetic and gravitational fields was strongly motivated by his desire to exorcize the demon of probability in quantum mechanics. In addition to believing that God was not a dice-throwing gambler, Einstein made another famous declaration regarding quantum mechanics, stating, "Subtle is the Lord, but malicious he is not." In any case, Einstein failed in his efforts to find a deterministic unified field theory that made sense to him.

Einstein's effort at a unification of gravity and electromagnetic forces was not the only attempt to do so. I became interested in separate attempts to unify gravity and electromagnetism by German-born mathematician and physicist Theodor Kaluza (1885–1954) and Swedish physicist Oskar Klein (1894–1977). Like Einstein, I had reasons for my interest in unified field theories: I was motivated by the possibility of using such a theory for time travel.

In 1921 Kaluza was able to show that by going to five dimensions a means could be found for combining the gravitational and electromagnetic forces. The fifth dimension was conceived by Kaluza as an extra dimension of space, but unlike the other three dimensions of space (length, width, and height) this extra dimension of space could not be directly measured. Kaluza was trying to combine the electromagnetic force with the gravitational force of Einstein's general theory of relativity. In Einstein's theory, the gravitational force is the bending of the three ordinary dimensions of space and the slowing down of the fourth dimension of time. Kaluza showed that, by combining his extra fifth dimension with the three dimensions of ordinary space and the fourth dimension of time, the electromagnetic course could be combined with the gravitational force. Since Klein also independently developed this theory, it came to be known as the Kaluza-Klein theory. At one point, Einstein studied the Kaluza-Klein theory but decided to try an approach that did not use a fifth dimension. The Kaluza-Klein theory is still studied by physicists today as a possible approach to a unified field theory.

The idea of unification goes back to the very foundation of physics. In fact, the first unified field theory was developed by the nineteenth-century Scottish natural philosopher (as theoretical physicists were then called) James Clerk Maxwell (1831–1879).

Maxwell's theory unified what was previously thought to be separate electric and magnetic fields. Maxwell was mathematically able to show that it was possible to use a changing electric field to produce and control a magnetic field. Earlier in the nineteenth century it had been shown by the English chemist and physicist Michael Faraday that a changing magnetic field could be used to produce an electric field.[15] Using Faraday's observations, Maxwell was able to mathematically synthesize his results with Faraday's observations to create a unified theory of the electromagnetic field. In this theory, the changing electric field produces a changing magnetic field. That changing magnetic field in turn produces a changing electric field and so on. The changing electric and magnetic fields spread out through space at the speed of light. Maxwell deduced that light was actually changing electric and magnetic fields. This electromagnetic theory of light was the first unified field theory. Maxwell's unification of the electromagnetic field has led to an unprecedented control of this force of nature. The fruits of modern technology, from electrical power generation to television, are a result of unification of the electric and magnetic fields.

One of the predictions of general relativity is that a clock in a strong gravitational field will run slower than a clock in a weak gravitational field. For example, suppose you have synchronized two identical atomic clocks.[16] If you were to take one of the clocks to the top of a high mountain where the gravitational pull of the earth is weaker, and compare it to the identical atomic clock at ground level where gravity is stronger, you would find that the clock at ground level is running slower than the clock at the top of the mountain. This connection between time and gravity fascinated me. I thought that perhaps a unified field theory could be used to control gravity and hence time. I decided to make a serious study of the modern unified field theories.

One of the major achievements of modern physics has been the unification of the electromagnetic force with the weak nuclear force. This unified theory is known as an "electroweak theory." The unification was finally achieved in the late 1960s by Steven Weinberg, Sheldon Glashow, and Abdus Salam, for which the three shared the 1979 Nobel Prize in Physics "for their contributions to the theory of the unified weak and electromagnetic interaction between elementary particles."

The electroweak theory led to the prediction of a new particle of nature. This new particle is a heavy photon. An ordinary photon has no charge and is essentially a particle of light with no mass. The new particle, called a Z particle, also has no charge, but it does have mass. These particles have been seen in the high-energy collision of elementary particles.

I read about earlier attempts that had been made to unify the strong, electromagnetic, and weak forces into a grand unified field theory. One of the predictions of the grand unified theory is that the proton, first discovered in 1918 by Ernest Rutherford,[17] which is a basic component of the atom, does not live forever as had been previously thought. Rather, it disintegrates after a sufficient time. Since this disintegration has never been seen, grand unified field theories have not been accepted.

I was intrigued by the electroweak theory and decided to see if I could develop a unified theory that would combine the gravitational force with the weak nuclear force. I thought I saw mathematical similarities between the two forces and that this might point the way to their unification, so I spent my days working for United Technologies and my evenings studying unified field theories that might be connected to gravity. I believed that if a complete unified field theory was found that combined all of the four basic forces then maybe one of the forces could be used to control the other, and perhaps this could lead to a working time machine.

The constant pressure of my day work and my late-night studies began to take their toll. I yearned to be teaching and developing my ideas in general relativity full time. I was twenty-eight years old and already feeling that time was wasting. I had heard it suggested that a scientist's best work is often done earlier in life, by the age of thirty, and I didn't want to miss my window of opportunity.[18]

Another opportunity Dorothy and I decided we did not want to miss was having children and becoming parents, as so many of our friends were doing. There came the day when she thought she was pregnant and went to the doctor. A phone call the next day confirmed what we hoped for. Within minutes we began telling the entire Western world. A month later, Dorothy awakened one morning with severe pain and profuse bleeding. She had lost the baby. We found it difficult to accept, but knew we still had a possibility of trying again.

After two years, I told my boss, Jerry Peterson, that I intended to leave and find a teaching and research position in a university. At first he thought that I was negotiating for more money, and he offered me a pay raise. When I told him that wasn't my intent, and how I felt more at home in academia, he was sympathetic. He also issued a warning: "You realize that in academia you're going to be taking an oath of poverty?"

When I assured him that I realized pay in academia would never rival that of private industry, he gave up and set out to help. Jerry knew Steven Weinberg, then at Harvard, and he sent the future Nobel Prize winner some of the independent research I had been doing in attempting to unify gravity and the weak nuclear interaction. Weinberg was kindly in his reply, saying that the model had some problems but the idea was interesting. Jerry also put me in touch with the head of the physics department at the University of Connecticut, Joseph Budnik, who had been doing consulting work at United Technologies during the summers. Budnik has an outgoing, bigger-than-life personality; I liked him immediately and knew it would be great to be part of his physics department. I was also fortunate in that Jerry's wife, Cynthia Peterson, was a faculty member in the UConn physics department.

I was invited to give a physics colloquium at UConn on my thesis research, which gave me the opportunity to meet the head of the elementary particle and fields group, Kurt Haller, who grew up in Austria and received his training in physics at Columbia University. He was friendly and encouraging, and, if I got a position in the department, I would be part of his group. Not long after my talk, I was offered a one-year provisional assistant professor position at the main campus of the University of Connecticut in Storrs. I was beside myself with joy, even though it meant a 50 percent cut in pay. Now, I could seriously return to my studies.

During my two-year stint in private industry, I had learned a lot about lasers and had a deeper appreciation for the interplay between theory and experiment in physics. Too, I was beholding to Jerry Peterson for assisting me in so many ways. I came to compare his managerial brilliance in handling a team of scientific free thinkers to the cohesiveness that Robert Oppenheimer had brought to the Manhattan Project. Nonetheless, at the time, I never said a word to Jerry about my longtime interest in time travel, as I was certain he would be skeptical. Ironically, my research days under his tutelage at United Technologies would forever change the way I did theoretical physics, and my work with lasers would one day end up providing a missing element to my design of an experimental time machine.

At a farewell party for me at the Research Center, the staff gave me a going-away gift: an authenticated signature of Albert Einstein. My youngest brother, Keith, who had become a successful professional artist living in San Diego, painted a brilliant portrait of Einstein to mount my authentic signature next to. I framed them together, and hung them on the wall of my study.

Under the gaze of the grand old relativist, I was ready to focus my attention on teaching, while probing further into the time-travel possibilities of general relativity.

Eight

Finding My Academic Home

⧗

Having traveled back in time, I am once again in the Bronx.

It is a warm Sunday evening in April, 1955.

Entering the building at 1455 Harrod Avenue, I take the lift to the eleventh floor. The hallway is narrower than I remember. I knock on the door of 11D.

When the door opens, my handsome father is standing before me.

I am surprised at how young he looks. My father is, in fact, only three years older than I, and I am now several inches taller.

"Can I help you?" His voice has a deep yet soft quality I remember.

I am not surprised that he doesn't recognize me. Although I have anticipated this moment for a long time, I am speechless when it arrives.

Boyd Mallett looks puzzled.

I finally find my voice. "I have some things to tell you that will be difficult to believe. But first, I want to show you some photographs."

One by one I hand him the old photographs of him with his wife and young children. Another image is of him working on a TV set. I know he had seen these pictures before because they are in our family album.

"Where did you get these?"

Without answering, I show him some photos he has not seen: of his children grown, and his wife—still beautiful, but years older.

"What is this all about?" He sounds alarmed.

I ask him if I can come in for a few minutes.

He seems to be considering my request. Then he stands back and opens the door fully, and leads me into the living room. I am surprised that I have remembered the layout so accurately, although it feels odd to be looking down on the room from this height. We sit down, and he again asks what is going on.

I gaze at this man who considers me a stranger—this man who, in both life and death, gave such unbelievable meaning to my own life.

"Look," I finally say, "you know about television. You know that by using electronics you can generate signals and transmit images across space."

He tilts his head to one side. He is interested.

I tell him that I have built a device that can send images and objects across vast periods of time. I explain that this device is called a time machine.

"A time machine? Is that how you got those pictures?"

"That's part of the story."

He wants to know more.

I knew matters of electronics and science would pique his interest.

"You should know that I built it because my father died." I tell him that my father worked and lived hard. "Then, in May 1955, on the evening my parents were celebrating their eleventh wedding anniversary, my father died suddenly of a heart attack."

He looks puzzled. "That's interesting. My wife and I will be celebrating our eleventh anniversary next month."

I go on—telling him that I was ten years old when my father died, and that I was so lost after his death that I didn't know what to do for a long time.

He seems to sadden. *Of course.* He knows about losing his father.

"Then I read this wonderful book called *The Time Machine,* by H.G. Wells. It became clear to me what I had to do. It gave me hope."

His facial expression suggests that he is beginning to put the pieces together. He looks hard at me. "You're not trying to tell me—"

"I built my own time machine, yes. I have traveled from the future to see you, Dad. I am your son Ronald."

"My son Ron is with his mother at a church function."

"Like every Sunday night. I know. That's why I came when I did, so we could be alone. I'm as real as that little boy at church with his mother. I am older and come from a different time, but I am Ronald Mallett. Please believe me. It's important that you listen to what I have come to tell you. You are in danger. You have a weak heart, Dad. You need to go to the doctor right away or you'll be dead in a month. And for God's sake, stop smoking."

Then I told him one of the most important things I came to say—made so because I can't remember telling him while he was alive. "I love you, Dad."

This is my fantasy. This is where I have gone so many times that I have memorized the script containing the words of my father and myself. I usually go there at night, in the dark, with my eyes shut, sweetly suspended between sleep and consciousness. I always return from the imagined visit wondering if he would really change his life. Could he? Could I alter his fate?

Might I save my father from his tragic, premature death?

* * *

On the morning of September 2, 1975, Dorothy kissed me good-bye and wished me good luck as I left our townhouse in Manchester. I drove to the Storrs campus of the University of Connecticut for the first day of school, and the beginning of my new life as an assistant professor of physics.

Storrs is situated in the northeast corner of Connecticut about twenty miles from the state capital of Hartford. The lush, tree-studded campus is a smaller scale version of the Penn State campus. To the left as I drove through the entrance were sprawling farmlands and, nearby, a rustic barn and some grazing livestock. Further down the way were the buildings of the College of Agriculture. The roots of the University of Connecticut, like those at Penn State, are firmly planted in soil agriculture. In 1881 UConn was established as Storrs Agricultural School on 170 acres of land donated to the state by brothers Charles and Augustus Storrs, successful New York businessmen who were born and raised on a Connecticut farm. By the time I came to UConn it had become the flagship university for the state of Connecticut.

Following written directions, I drove into the center of campus and turned on North Eagleville Road toward the physics building, a modern structure with a large dome on the top that housed a telescope. I parked, entered the building, and took the elevator.

I felt a bit anxious as I walked into the office of the physics department. At thirty, I was no more than ten years older than the average undergraduate, and only a year or two older than many graduate students. There was also the issue of race, as I was the first and only black faculty member of the physics department. Nonetheless, I had a very specific image of what a college professor should look like right down to the white shirt with a button-down collar and the tweed jacket with elbow patches. If I was to be taken seriously, I decided that I was going to have to walk the walk. My mother's influence and personal pride were remembered—how even when she was working as a store maid she had dressed for success. I was wearing a button-down shirt and appropriate tweed jacket with leather patches.

The physics department head, Joe Budnick, gave me a warm welcome. He took me around for introductions to the office staff. They were friendly and helped put me further at ease. I was assigned to room P-414 on the fourth floor. My office was spacious but without much of a view. From my window I looked directly out at the Life Science Building, with a partial view of the campus cemetery, where, I would learn, a number of retired professors were buried. After settling in, I went around to visit my faculty colleagues—those who were still alive, that is.

In the physics department at the time were some thirty faculty members engaged in all areas of modern physics. Their various fields were broadly represented by a number of defined groups: atomic and molecular, particles and fields, condensed matter, and nuclear. My area of research, Einstein's general theory of relativity, came under the particles and fields group nominally headed by Professor Kurt Haller, whom I had met when I gave a lecture on campus while still working for United Technologies.

I reported to Haller's office located at the end of the hall on the fourth floor. Haller was a genial man of slight build and medium height with a certain Old World aura befitting his early years in Vienna. When he was ten, in 1938, he and his family had fled Austria after the Nazi annexation of their country. Kurt received his Ph.D. in theoretical physics from Columbia in 1958 and six years later joined the UConn physics faculty. He had done some important fundamental work in theoretical elementary particle physics, specifically in an area known as gauge theories, which have their roots in the ordinary practice of measurement.

To gauge something means, of course, to measure it according to some kind of scale. Consider identical twins of the same height. Suppose one twin lives in Boston and the other in New York City. On a particular day they decide to measure their height. The Boston twin has a common yardstick that measures distance in feet and inches, while the New York twin has a meter stick that measures meters and centimeters. The Boston twin measures his height as six feet, and the New York twin measures his height as 1.8 meters. The measurements sound very different, but clearly the height of the twins hasn't changed. Rather, it has to do with the scale by which they were measured. In this case, there is a rule called a conversion factor that tells you how to go back and forth between meters and feet. The conversion factor is .3 meters divided by one foot. That is, if you multiply six feet by .3 meters per foot you get 1.8 meters. This is what gauge theories in physics are all about— the notion being that the forces of nature are conversion factors needed to compensate for changes in scale.

For example, in quantum mechanics an electron behaves like a wave. If we start with two electrons we can change the scale of the wave of one of the electrons by shifting its wave relative to the other electron. The shifting of the wave leads to a compensating conversion factor that we call the electric force between the electrons. In other words, the two electrons feel an electric force between them to compensate for a shift in the electron waves.

Kurt Haller, a leading expert in gauge theories, was interested in my research because it was connected with the idea that gravity was a gauge theory. A way to think about this is as follows: Suppose you are unfortunate enough to be in a glass elevator that is falling freely. Surprising as it might seem, since you are falling at the same rate at the elevator is falling, you would find yourself suspended in space in the elevator. In other words, you would feel no force. However, as you look outside and see the floors rushing by, you would know that something is pulling you to the ground. Your changes in scale or gauge (i.e., the change in the floors as you drop) lead you to conclude that there is a conversion factor called gravity that will probably make the end of your ride an unpleasant one. In other words, gravity is detected as a difference in acceleration between different locations. This is the gauge theory of gravity.

From that first day I walked into his office as the department's newest faculty member, Kurt encouraged me to continue with my own research. I was happy to oblige, while being careful not to mention that I was interested in gravity for other reasons—namely, the part gravity might play in time travel.

My reason for keeping mum about my interest in time travel was one of practicality: to advance up the academic ladder, from assistant to associate to full professor, and in order to eventually attain tenure (meaning one cannot be dismissed without cause), it is best not to be pegged within or outside your department as a crackpot. While in physics creativity and strange ideas are often encouraged—black holes are very strange and by definition invisible objects, but research on them has long been considered legitimate—time travel was deemed in the 1970s as too extreme a subject for serious scientific inquiry. My plan was simple: as I progressed in my academic career toward tenure and full professorship, which would provide me with both financial security and professional acceptance, my time travel work would be my own business. Afterward, when I was a tenured full professor, I would feel freer to publicly work on, and speak out about, the subject that had caused me to go to college, study hard, and become a physicist in the first place.

As I visited with other faculty, I got more insight into survival techniques in academia. I was, in fact, on probation. Academic probation for a faculty member meant a year-by-year appointment. This probationary period continued until one achieved tenure, which usually took a number of years. Officially, there are three components to the responsibilities of a faculty member. These are teaching, research, and service to the university community. Unofficially, one department veteran took me aside and said, "Ron, in reality the three aspects of your job are research,

research, and research." I took this advice to heart, although I knew from other faculty members that being a good teacher was important, too.

My teaching assignments included undergraduate and graduate courses. I taught undergraduate general physics, an algebra-based course that covered electricity and magnetism, optics, and modern physics. My graduate course was Methods of Theoretical Physics, covering vector analysis, matrix theory, ordinary and partial differential equations, and group theory.

There is a truism to the effect that the best way to learn a subject is to teach it. It is not mysterious how this happens; to be a good teacher one needs to prepare, and adequate preparation requires much study and thought when starting from scratch with no course outline or notes. I spent as many as five or six hours in preparation for an hour-long lecture. A graduate class took longer—up to eight hours preparation per classroom hour. My efforts paid early dividends: in my first year, I was surprised by a spontaneous ovation after delivering a lecture on electricity and magnetism to my general physics class. I am not sure any compliment, before or since, has meant quite as much as that one.

Near the start of that first term, I was advised by a senior member of my department that it would be wise for me to publish a "solo paper" right away. He acknowledged that I had published two previous papers with my thesis advisor, but as a theorist in my first faculty position, he said it was vital for me to do my own paper to prove that I could work independently. After that, he went on, I should work on publishing a paper with a graduate student, which would show that I could also supervise student research work.

As a possible subject for my first solo paper, I ruminated more about gravity as a gauge theory. I eventually decided there might be an interesting and new connection between gravity and the

Einstein-Infeld-Hoffman (EIH) problem, about which I had lectured in front of Professor Fleming and others while a graduate student at Penn State. I delved headlong into the subject, delighted to have the hours during the day to do research, rather than having to do so on my own time after a full day's work as I always had before.

In the original EIH problem, Einstein and his collaborators were able to show that the equations for gravity were all that was required to determine the equations for motion of matter. That is, if you have a gravitational field you can determine the equations that tell you how matter will move in that gravitational field. My calculations indicated that there was an inverse EIH problem—that with only the equations for motion of matter, the gravitational field can be determined. By using a gauge theory of gravity, I could show that changing the scale of the equations that determined the motion of matter would lead to the correct equations for Newton's laws of gravity. Amazingly, no one had attempted to turn the EIH problem around. I wrote up my results and submitted a paper to the respected physics journal *Physical Review,* which had published my second graduate paper. My first solo paper, "Symmetry Breaking and the Gravitational Field," was published in May, 1976.

With the publication of that article, I successfully survived my first year in academia as a probationary faculty member. As I was beginning to feel at home in the physics department, Dorothy was doing well herself, enjoying her work as an executive secretary with Travelers Insurance Company. That summer, she and I drove to Altoona for a visit. One sunny morning we walked along the old railroad tracks that bisect the town. I had come this way so often as a boy that I began reflecting aloud on some of my childhood experiences, both good and bad. Suddenly, I came to an abrupt halt as something occurred to me that I hadn't thought of in years. I turned to Dorothy and explained that one day when I was about fifteen years old I had taken this exact route home from school and made a "solemn promise to myself."

"What was the promise?" she asked.

"I stopped right about here, and swore to myself that when I grew up and finished building my working time machine, I would transport myself back here to that day—and tell my younger self that I had succeeded."

Dorothy smiled. "I take it that hasn't happened."

"No," I said, returning her smile. "Not yet."

It was a fanciful idea, granted.

Yet the dream, still hidden from all but those closest to me, lived on.

My mother and father, Dorothy and Boyd Mallett, with me (left) and my baby brother, Jason, at the Bronx Park, 1948.

My soldier father before being shipped overseas in 1944.

Visiting my grandparents' farm in Claysburg, Pennsylvania, with my mother and brother Jason (left), early 1950s.

Dad working on a television set, 1954.

The cover of the Classics Illustrated edition of *The Time Machine,* which I brought for fifteen cents the year after my father died. The story gave me hope of one day building a working machine that would take me back in time to see Dad again.

First page of the Classics Illustrated edition of *The Time Machine.* I used such illustrations as an elementary blueprint to build my first time machine at age eleven.

My graduation picture from Altoona Area High School, class of 1962. Withdrawn and having few friends, I wanted desperately to leave home and be on my own. Within two weeks of graduation, I shipped out for U.S. Air Force recruit training.

Dorothy Fry and me on the steps of our apartment in Brooklyn, 1968. At the time, I was working as a research technician at the Markite Corporation in Greenwich Village. We were married for more than twenty years.

In my office in the University of Connecticut Department of Physics in 1989, two years after becoming a full professor. Today, the poster of an old friend and longtime inspiration is still taped onto the same filing cabinet not far from my desk.

I served for six years as a campus liaison officer for the U.S. Navy Reserve. On this day in 1982, I went on a training cruise aboard USS Miller (FF-1091), a fast frigate named in honor of Dorie Miller, an African-American ship's cook who manned a machine gun—a weapon he had not been trained to operate— aboard his docked ship at Pearl Harbor on December 7, 1941. Miller, who succeeded in downing one of the attacking Japanese aircraft, was awarded the coveted Navy Cross.

At the International Center for Theoretical Physics in Trieste, Italy, in 1979 with physicist Joseph Taylor and his wife, Marietta. A year earlier, Taylor's breakthrough research on binary pulsars with graduate student Russell Hulse had been announced, providing the first experimental evidence for the existence of gravitational waves. For their work, Taylor and Hulse shared the 1993 Nobel Prize in Physics.

Visiting legendary physicist John A. Wheeler at his home on High Island, Maine, 1995. I spent my first sabbatical (1982–83) as a Ford Foundation fellow at the University of Texas, Austin, where Wheeler was the director of the Center for Theoretical Physics. Now ninety-five years old, Wheeler still keeps an office at Princeton.

Boyd Mallett's three sons (from left): Jason, a retired sales executive; Keith, a successful commercial artist, and me. Circa late 1990s.

In my office explaining to graduate students the mathematics and science behind my laser-driven time machine, under the watchful gaze of the old relativist, 2006.

Dr. Ronald Mallett, professor of physics, University of Connecticut.

All photos from the personal collection of Ronald Mallett.

Nine

My Expanding Universe

⧗

With the calendar showing the approach of fall, I could hardly wait for classes to resume. I was particularly enthused to be scheduled for a general relativity graduate course, which would cover tensor analysis, differential and Riemannian geometry, and, importantly, Einstein's gravitational field equations, which show how a massive body like the sun causes empty space to curve. This curvature of space results in what we call gravity. In truth, the planets of our solar system are not influenced so much by gravity as they are pulled through this curved space created by the sun. It is basic, universe-building stuff that I knew could be taught entertainingly with experiments as well as with equations.

The class ended up with only twelve students, which I would learn was a decent number for a high-level graduate physics course. It was an older and rather sedate group, unlike the eager-beaver undergraduates I had taught the previous year. Nonetheless, I was excited to be leading these dedicated and focused students through this challenging course about some of the most vital elements of modern physics, with a sizable dose of

Einstein stirred into the mix. It still felt unreal that I was actually a physics faculty member at a major university, with the opportunity to teach, of all things, *general relativity*.

During the term, I managed to slip in a lesson from Gödel's study of cosmology, which I first read about in the air force. Although there was much I didn't understood the first time around, I had become instantly fascinated by Gödel's work because it allowed for the possibility of time travel into the past.

Cosmologists frequently study theoretical model universes that may be similar or different from our actual universe. They study these model universes just to see what might be possible within the laws of physics. In his rotating-universe work, Gödel allowed for closed loops in time, or closed timelines. Normal time is represented as a straight line, the bottom of the line representing the past, the middle of the line the present, and the top of the line the future. Such a timeline shows the world as we normally experience it; that is, we move from yesterday, to today, to tomorrow. A closed timeline means that the ends of the line are connected. That meant, Gödel concluded, that travel into the future or the past is equally possible along a closed loop line.

Preparing for the general relativity course allowed me the opportunity to delve deeper into circular timelines. This came about in connection with the theory of black holes. Previously, I had only studied black holes that did not rotate. Real stars rotate just like the earth. Black holes are formed from the gravitational collapse of stars. If the rotation of the resulting black hole is slow enough, one can neglect the gravitational effect of the rotation. For a rapidly rotating black hole the situation becomes very complicated. In fact, it took nearly fifty years before a solution of Einstein's field equations could be found for a rotating black hole, a solution found in 1963 by New Zealand mathematician Roy P. Kerr.

Wanting to know more about the Kerr solution, as the gravitational field for a rotating black hole came to be known, I found a published paper, "Global Structure of the Kerr Family of Gravitational Fields," by physicist Brandon Carter (born 1942). The abstract of Carter's paper caught my attention when I read his statement that a rotating black hole has "closed timelike lines which are not removable." Closed timelike lines are just another name for circular timelines, and are the same timelines that appear in Gödel's theory of the rotating universe. This meant that there was some connection between the work of Carter and Gödel. As I studied Carter's full paper and worked his calculations, I became increasingly intrigued. He found that the closed timeline of a rotating black hole would allow travel into the past. In other words, a rotating black hole could be used—theoretically, at least—as a kind of transport tunnel into the past. I decided then and there that I would learn more about rotating black holes. Such research, and even publishing my findings, would allow me to study how time is affected by gravity while providing cover for my primary interest in figuring out how to build a time machine.

As I was giving serious consideration to the advice that had been proffered my first day on campus regarding finding a graduate student with whom I could work on publishing a paper, Fred Su, a Chinese-American graduate student who wanted to conduct original research in general relativity, came knocking on my door in search of a thesis advisor. We kicked around some possible topics before it dawned on me.

"Rotating black holes are very interesting, Fred," I said. "These are also known as Kerr black holes. They have some really strange properties and one of the strangest is the fact that they affect time. Rotating black holes can actually lead to closed loops in time. That means you can go back into the past in a rotating black hole."

Actually, time wasn't the only thing about black holes that interested me.

They also affect space, and one of those effects is called frame dragging, which can be defined as the stirring up of space. It had been known for a long time by general relativists that a rotating mass (such as a rotating black hole or planet) drags space around in its invisible wake. Frame dragging looks rather like twirling an apple (mass) round and round into a bowl of caramel sauce (space).

When explaining the nature of black holes to my students, I always start by telling them how stars are formed, a process that commences when the heavy-hydrogen (also called deuterium) gas atoms in interstellar space start attracting one another gravitationally, not unlike a Friday night crowd at a singles' bar. The nucleus, or core, of a heavy-hydrogen atom consists of a proton plus a neutron. As these heavy-hydrogen atoms collect together, they gravitationally attract more atoms. When two heavy-hydrogen atoms collide, they combine to form a helium atom, whose nucleus is composed of two protons plus two neutrons. If the mass of the two original heavy-hydrogen atoms was measured and compared with the mass of the resulting helium there would be a discrepancy, with the helium atom being something less than the two original heavy-hydrogen atoms. This loss of mass gets turned into energy via $E = mc^2$. It is through this combustible process of heavy-hydrogen gas combining to form helium that a new star is born.

Every star, including our sun, exists thanks to a balancing act between two opposing forces. The gravitational attraction of the gases of the star constantly try to pull the star inward, while the heat radiation from the energy given off by the colliding heavy-hydrogen atoms continually pushes outward. This push-pull process goes on until the internal fuel provided by the heavy-hydrogen gas is used up to form helium. As this point, there is no more fuel to feed the internal heat and pressure necessary to keep the star pushing outward, and the gravity pulling the star inward is no longer kept in check. As a result, the star begins to collapse. As it does, a number of interesting things happen.

First, the star's electrons (lightly charged subatomic particles that make up less than one percent of an atom's total mass) get pushed together, also by gravitational attraction. There is a law in quantum mechanics called the Pauli exclusion principle, developed by Swiss physicist Wolfgang Pauli (1900–1958), which states that two electrons cannot occupy the exact same space. This principle keeps the electrons from being pushed together. The gases will eventually settle down, turning the mass into a white dwarf, a star about the size of earth that has exhausted its nuclear fuel and cannot produce heat to counteract the force of its own gravity. This will be the fate of our five-billion-year-old sun, considered a middle-aged star, when it burns out in another five billion years.

However, there are exceptions. If the star is significantly larger than our sun, the electrons will get pushed into the protons by the gravitational attraction to form neutrons and become a neutron star which is a dense star (technically a stellar corpse) composed mainly of neutrons. For stars of mass greater than two and a half times the mass of our sun there is nothing that can keep them from continuing to collapse inwardly. As they do, the gravitational field continues to escalate. Eventually, the field of gravity around such a star will be so strong that anything that tries to leave the surface of the star, including light, will get pulled back. With no light able to leave the surface, the star becomes an object that was first coined a "black hole" in the 1960s by American physicist John A. Wheeler.[19]

Fred was intrigued with the idea of studying black holes, but I knew I had to be cautious when it came to directing his research work. I certainly couldn't assign him the task of researching time travel into the past via rotating black holes, or we would both pay a price. A student thesis needs to be solvable as well as specific and original, so I suggested finding a thesis topic in an arena that I had been studying in electromagnetism called the Faraday effect (named after the English physicist and chemist Michael Faraday), which has to do with what happens when light is sent through a transparent material (like glass) in a magnetic field. As light moves forward it oscillates up and down like a wave on a string. The up-and-down direction of the oscillation is called the polarization of light. (Incidentally, polarized sunglasses make use of this property of light. When light from the sun is reflected off a metal surface—like the hood of a car—it becomes polarized. A polarized lens blocks only the polarized light.)

When a light wave is sent through a glass block, and when a magnetic field is applied to the block in the direction the light wave is moving, the plane of polarization is twisted, or rotated. I had read papers suggesting that the gravitational field of a rotating body would cause a twisting of the plane of polarization of light as it passed in the direction of the axis of the rotating body. Since a rotating black hole causes a twisting of space, this implies that it would also cause a rotation of the plane of polarization of light going by it, although this particular effect had never before been proven. I suggested to Fred that his thesis topic could be to look at what happens to polarized light when it goes past a rotating black hole.

Fred was eager to begin his research, although not before informing me that if he succeeded in breaking new ground, he wanted us to submit his paper to one of the premier refereed journals for publication. I smiled inwardly. *Where had I heard that before?* I asked if he had one in mind. He said the *Astrophysical Journal,* which I knew was one of the most difficult journals in which to get published. "Okay, Fred," I said. "Let's go for it."

The third official component of my job—university service —soon reared its head. I understood there was a role for me to play as it related to the community at large, and was particularly keen about giving public lectures and being involved in minority affairs. As the only African-American member of the UConn physics department, I relished recruiting and advising minority students interested in studying the sciences, as there was a dearth of representation in the booming fields of science and engineering. But as far as any interest or involvement in university administrative activities or campus politics, well, that was another matter. After a frank talk about faculty administrative responsibilities with an assistant vice president of academic affairs, Joan Geeter, who ended up becoming a friend, we agreed early on that she would take care of the university, and I would take care of the universe.

Ten

Deeper into Black Holes

⏳

The year 1979 would usher in worldwide events to celebrate the centennial anniversary of the birth of Albert Einstein, who was born on March 14, 1879, in Ulm, located on the left bank of the Danube in southern Germany.

One major event recognizing Einstein's pivotal work was to take place in the summer at the International Center for Theoretical Physics in Trieste, Italy. Hoping to somehow be part of the activities, I sent an application to the organizing committee more than a year in advance, proposing to give a talk on the inverse Einstein-Infeld-Hoffman problem, about which I had, by then, published a number of scientific papers. I made further inquiries, and found that in the event I was invited to give a paper at the conference the university would provide partial funds for my airline ticket and some hotel expenses. However, I would have to make up the difference out of my own pocket.

Financial salvation came in the form of a consulting job with Pratt & Whitney Aircraft. Someone there remembered my work for United Technologies, specifically, my theoretical analysis of laser-hole drilling. Now, they wanted me to conduct the same type of research using electron beams to drill holes. In addition, I was asked to teach one of the staff engineers the mathematical techniques I had developed for laser-hole drilling. This moonlighting in the summer of 1978 when school was not in session worked out well, and I was able to save enough money to pay for the rest of the trip.

My application to the Einstein conference, which was to be held at a resort near the Adriatic Sea, was accepted in early 1979. Dorothy and I—neither of us had been to Europe before—planned to make a vacation of it, too. A week before we were to depart, some friends held a going-away dinner party for us.

We left the party shortly after midnight, with Dorothy driving. As we entered a turn on a rural road, the high-beam headlights of an oncoming car blinded us. Dorothy steered over to the shoulder as far as possible, coming to a stop at a grassy embankment. A second later, the other car collided with us. I had only minor cuts and bruises, as Dorothy's side of the car took most of the impact. (The other driver turned out to be drunk, and barely had a scratch.) When I looked over at Dorothy, I could see that the driver's door and dashboard had her pinned. She was lying like a rag doll in a frightful position. Fireman had to use metal-cutting tools to get her out, and an ambulance took us to the hospital. Dorothy had a badly fractured hip and a broken leg. The orthopedic surgeon on call was summoned, and Dorothy was wheeled into emergency surgery. In a bizarre coincidence, the surgeon, Doug Griswold, had been our host that evening—along with his wife, Pat Terry, from the UConn English department. Doug repaired Dorothy's hip with pins and a plate, and inserted a stainless steel rod to support her broken leg. A full recovery was expected, but she would be in the hospital for three weeks.

One thing was certain: Dorothy would not be traveling any time soon. Our trip to Italy would have to be cancelled. The second day I visited her in the hospital, she made a case for me to go to Trieste without her. She explained that she was getting good care, there was not much I could do for her at present, and I would be back before she was released from the hospital.

After arranging for friends to visit Dorothy during my absence, I reworked my itinerary, trimming the extra vacation days, finished packing, and flew to Trieste to keep my date with Einstein at his hundredth birthday bash.

* * *

Trieste is an extraordinarily beautiful city set in the isolated northeastern corner of Italy that has at times been occupied by Austria and Yugoslavia.

About 300 physicists from throughout the world and a variety of disciplines were in attendance at the conference. I was scheduled to deliver my presentation midway through the schedule, and I spent most evenings after dinner sitting on the deck of my hotel room overlooking the azure waters of the Adriatic going over my paper for the umpteenth time. About twenty people attended the evening workshop where I spoke, and afterward there was a lively question and answer period.

The next day, relieved of my conference responsibilities, I took a late-afternoon stroll through picturesque Trieste, which was rapidly becoming the most important center for scientific research in Italy. Working up an appetite, I stopped for Wiener schnitzel, and happened upon the best I had ever tasted (before or since). Raving about it the next day to my colleagues, one of them said the excellent Wiener schnitzel here was the result of the Austrian influence. When it comes to science and good food, it seems we physicists know our stuff.

I made a valued friendship in Trieste. University of Massachusetts astrophysicist Joseph Taylor is a tall, thin man who in posture and physique puts me in mind of the actor Jimmy Stewart. A friendly man with a quick smile, Taylor and I spent a lot of time talking about science and life, and enjoying meals together. In a serendipitous blend of friendship and work, Taylor's breakthrough research, for which he would eventually win a Nobel Prize, would deepen my own understanding of the flexibility of spacetime.

In 1974, Taylor, while a professor at Amherst, and his graduate student, Russell Hulse, had utilized the world's largest single-dish radio telescope at Arecibo, Puerto Rico, to discover a pulsar (a rapidly spinning neutron star) emitting radio pulses at intervals that varied in a regular pattern, decreasing and increasing over an eight-hour period. They concluded from these signals that the pulsar had to be alternately moving toward and away from Earth. They deduced that it had to be orbiting around a companion star, which they believed was also a neutron star.

One of the key predictions of Einstein's general theory of relativity is that two stars (binary stars) in orbit around each other will set off ripples in spacetime. Since gravity, according to Einstein, is the curvature of spacetime, then these ripples of spacetime would be waves of gravity. Gravity waves had not been directly detected before.

Taylor and Hulse observed for several years the decay of binary pulsars, and went where others before them had not. Their breakthrough finding, reported in 1978—the year before I met Joe in Italy—gave the world the first experimental evidence for the existence of gravitational waves—strong support for Einstein's theory of gravity.[20]

As much as I liked Joe from the moment we met, and found him to be open and unpretentious, I did not mention anything about wanting to build a time machine. Not only was I still concerned about the crackpot issue and my advancement in academia, I also had come to realize that I faced another problem: it was difficult enough to be accepted seriously as a black physicist (I was the only one in attendance at Trieste) without being associated with what was still a fringe area of research.

Before we departed Trieste, I invited Joe to give a physics colloquium at UConn on his groundbreaking work in astrophysics, which he subsequently did.

Back home, with Dorothy on the mend, I was eager to return to school that fall. During that school year (1979–80), an issue took the forefront that was more inevitable than surprising. I had long been bothered by the scarcity of minority students considering scientific and engineering careers. Indeed, attending the Trieste conference and finding myself the only black physicist had been a vivid reminder of the sorry state of affairs when it came to recruiting minorities for the sciences. At my own university, in fact, there were no minority physics majors and only a handful of engineering majors.

One of UConn's minority engineering students, Barry Walker, was starting a campus chapter of the National Society of Black Engineers. He approached me that fall about serving as a volunteer faculty adviser for the group. I was delighted to do so, and served for three years (1979–1982). The group began attracting larger numbers of minority students who had aptitude in math and science but had not seriously considered a career in engineering.

One activity I arranged for the chapter was a visit to the submarine base at Groton, Connecticut, which made for an exciting day. In the process, I became friendly with the local navy recruiters who arranged the trip. We discussed the difficulties that minority students had in the navy. As it turned out, the officer corps of the United States Navy at that time was extremely underrepresented by minority officers, which I was told led to escalating tensions within the ranks. The navy sought to turn the situation around by developing a new program called Campus Liaison Officer. The idea was to approach college and university faculty members who had

had previous military experience. The navy would offer these individuals a commission in the U.S. Navy Reserve with the rank of lieutenant commander in exchange for helping the navy increase its officer corps with scientifically and technically trained minority personnel. For me, after my service as an enlisted man in the air force, the offer was irresistible. The prospect of receiving a reserve commission and at the same time helping minority students seemed too good to be true. In a campus ceremony, a rear admiral commissioned me with the president of the university as well as a number of my students in attendance.

I spent six years (1979–1985) as a Campus Liaison Officer. One of my responsibilities was to serve as a role model for minority military recruits, as well as students. In this capacity, I was asked by an air force recruiter to visit Lakeland Air Force Base in San Antonio, Texas. Lakeland, of course, is where I had spent a "season in hell" years ago going through basic training as a lowly enlisted man. At the request of the air force recruiter, I visited the base in uniform. We went on a tour of a training area where a training instructor (TI) was yelling at a group of hapless recruits. The TI had his back to us. The recruiter quietly asked me a question. As I was responding, the TI suddenly said in a booming voice, "*Who* is that talking behind my *back!*" He spun around, saw all my gold braid, came to attention, and saluted. "Excuse me, *sir!*"

I returned his salute. "Carry on, sergeant."

As the TI turned his attention to the recruits, I thought back to the day I had gotten off the bus at Lakeland and been yelled at myself. I knew that the respect I had so demonstratively received from the TI for being an officer would not be lost on the minority recruits who were being yelled at now.

Meanwhile, I was quite pleased with the thesis work of my graduate student, Fred Su, who had succeeded in showing that polarized light (the up and down oscillations) from a star that passes a rotating black hole would get dragged or twisted around the rotating black hole. In coming up with his new result, Fred significantly advanced the understanding of how rotating black holes can affect space. As Fred had requested when he first started his work, we submitted his findings for publication in the *Astrophysical Journal*. The paper was accepted for publication in late 1979. In June 1980, the article appeared under the title, "The Effect of the Kerr Metric on the Plane of Polarization of an Electromagnetic Wave." Fred was ecstatic, as I well remembered being with the publication of my first paper at Penn State.

As for me, I had learned more about rotating black holes in supervising Fred's thesis. The thesis had not dealt with the time travel aspects of rotating black holes, but it did help me better understand the phenomena of frame dragging in connection with rotating matter. As a result, I was able to ask this question: Is there a connection between frame dragging and closed time-like lines of the Kerr rotating black hole?

If frame dragging in space did take place—as yet unproven—then the conditions for closed time loops might also exist. This meant that frame dragging might ultimately be connected with time travel to the past. That such a connection might exist was just a hunch at that point, and it came without one shred of scientific evidence. Yet, I decided that this avenue of inquiry was worth further investigation.

As I was thinking more about how closed time loops might facilitate travel to the past, *Somewhere in Time,* a beautiful and poignant movie about time travel, was released. Based on the science fiction novel *Bid Time Return,* by Richard Matheson (a frequent contributor to *The Twilight Zone*), the movie opens with a young playwright named Richard Collier (played by the late Christopher Reeve) at a party celebrating the successful opening of his play. Collier is approached by an elderly woman who gives him a pocket watch. "Please come back to me," she says before disappearing into the crowd. The woman is a complete stranger to him, and Collier is perplexed.

Some years later, Collier visits an old grand hotel and sees a portrait of a beautiful young woman. He learns from the hotel staff that the painting is of an actress named Elise McKenna (played by Jane Seymour), who was considered one of the finest actresses in the early 1900s. Collier becomes infatuated with McKenna's portrait, and visits her home. He is astonished to discover that she is the elderly woman who had given him the pocket watch. Unfortunately, McKenna has since passed away. At McKenna's house, Collier comes across a book entitled *Travels Through Time,* written by a philosophy professor who has studied time travel. Collier finds the professor, and learns that time travel might be possible using self-hypnosis in a suitably isolated environment linked to a previous time.

Collier returns to the old hotel and is able to project himself back to 1912 when McKenna was acting in a play at the hotel. The transition is dramatic, with the sight and sound of horse-drawn carriages and other period trappings. Collier meets McKenna and they fall in love. He shows her the pocket watch that she gave him. Accidentally, a coin falls out of his pocket which has a future date on it. Suddenly, Collier is unable to remain in the past and is returned to his future time. All that remains with McKenna in the past is the pocket watch. It is clear that McKenna will eventually see Collier again in the future and give him the watch back, which will enable him to come back to her, thereby completing the closed time loop. The storyline had more to do with psychology than physics, and yet the questions it posed and the possibilities it suggested were intriguing.

The movie reminded me how important science fiction had been in my life, allowing me at a young age, when my world was in such shambles, to dream big. I soon found a fellow sci-fi fan in UConn associate dean Karl Hakmiller, who would eventually become dean of the graduate school. We got to know each other while driving together to an educational conference at Harvard. As we headed for Boston, we discovered our shared passion and started excitedly discussing favorite books, stories, and movies. We ended up skipping the evening session at Harvard and going to Karl's favorite German restaurant in Cambridge to continue our lively discussion, which soon segued into time travel.

One short story, "Vintage Season," by C. L. Moore, was a mutual favorite. The story opens at a seaside inn, where the innkeeper overhears a group of new and rather strange guests talking about how this was one of the most beautiful times of this century. They seem to be talking about the present as if it has already happened. The innkeeper learns that these people are from the future, and they know that there is going to be a disaster in the area: a massive meteor shower is about to hit. These time travelers have come from the future to observe this disaster, something they routinely do. They observe but do not interfere; in a real sense, they are time tourists.

Karl was so animated in these discussions—particularly about time travel—that over coffee I made a decision. I told him about my father's death. Believing I could trust him, I then took the final step and told him about my lifelong dream to build a time machine to travel back to my father's time. "It's why I studied math and science, and why I ended up becoming a physicist."

Karl was familiar with the general ideas of Einstein's special theory of relativity, and we discussed how these might allow for time travel. In particular, he knew that Einstein had predicted that time slows down for a moving clock. And he knew that this led to the so-called twin paradox in which the twin that was moving close to the speed of light would age much less than his stay-at-home twin. In effect, the moving twin would have traveled into his brother's distant future. Karl understood why at this point in my academic career I wished to keep my ultimate goal under wraps. He seemed so moved by my story that I knew my instincts about him were right, and over the years they would prove to be so again and again.

Karl and I ended up becoming close friends as well as co-faculty advisors for the university's science fiction club. One evening a month, we would meet with about a dozen students in a room at the student union building to discuss sci-fi stories and books, and watch movies on one of the earliest versions of a videotape recorder, which I lugged back and forth from home to school. In such a setting we watched, *A Boy and His Dog,* a post-apocalyptic story; *The Man Who Fell to Earth,* starring David Bowie as an alien who comes to Earth; and *The Final Countdown,* in which a modern aircraft carrier goes through a time warp and ends up in 1941 just before the attack on Pearl Harbor, with the moral dilemma being whether they interject their modern-day firepower to stop the Japanese attackers and thereby change history. One novel we discussed was *Bring the Jubilee* by Ward Moore (1953), in which the South has won the Civil War. Thankfully, a historian is able to travel back into the past and change what happened at the Battle of Gettysburg to cause the future with which we are familiar: the North being victorious. I say "thankfully" because the idea of the South winning—and what, *preserving slavery?*—held some awfully unsettling implications for me.

Of course, the most successful time travel movie ever made, and one which spawned numerous sequels, was *Back to the Future,* released in 1985. The fact that teenager Marty McFly (Michael J. Fox) is transported by his brilliant but slightly mad inventor friend, Doc Emmett Brown (Christopher Lloyd), back to 1955 held special meaning for me. That was, of course, the year my father died. McFly nearly erases his own existence by interfering with the budding romance of two teenagers who are destined to become his parents. The movie dramatically illustrates a potential serious problem with any time traveling, that of interfering with the past.

I continued to publish my own research papers, while enjoying my teaching assignments. My peers seemed satisfied with my research, teaching, and community-based activities, and I was awarded tenure in March, 1980. At the same time, I was promoted to associate professor of physics, with only one more rung remaining to reach full professor. In order to achieve the latter, my ongoing research would be reviewed and evaluated not only by my own departmental colleagues but also by physics professors at other universities. I would need to keep on track in order to take that final professional step; many faculty members never reach the level of full professor their entire careers.

While progressing in my academic career, I had managed to increase my knowledge about relativity and the nature of time, but I was not any closer to understanding how to build a time machine. Then, out of the blue, an opportunity presented itself that might nudge me closer to my ultimate goal. I saw a posting on the physics department bulletin board soliciting candidates for a Ford Foundation Senior Postdoctoral Fellowship. This enabled those selected to study full-time for an entire year with anyone in their field.

I would be eligible for my first sabbatical leave at the university for the school year 1982–83. A sabbatical is an old tradition in which a faculty member is granted leave every seventh year for travel and research. This allows a faculty member to visit other institutions and collaborate with researchers in their field without the normal responsibilities of teaching and university service. At UConn, a sabbatical leave is granted for a year at half salary, which could cause some financial hardship. However, the Ford Fellowship would make up the difference.

I thought about who in the world of physics I would most want to work with. It did not take long for me to identify my first choice: John A. Wheeler, a legend in his own time. I had admired Wheeler's work from afar ever since reading his nontechnical article, "The Dynamics of Spacetime," while in the air force. Since then, I had read many more of his scientific papers.

Wheeler received his Ph.D. from Johns Hopkins University in 1933. As a National Research Council fellow, he was mentored by the great quantum pioneer, Niels Bohr (1885–1962). Later, working with Bohr, Wheeler helped to develop a theory of nuclear fission (1939). He went on to become a faculty member at Princeton University and a colleague of Einstein's. (Need I say more?) He also worked with J. Robert Oppenheimer. In the 1940s, Wheeler and his student, Richard Feynman, developed a formulation of electromagnetism. This work ultimately led Feynman to his Nobel Prize–winning contributions in quantum electrodynamics. By the 1950s, Wheeler had become increasingly interested in using Einstein's general theory of relativity to study stellar evolution. This work led to Wheeler's coining the term "black hole" to describe the ultimate fate of a dying star. He had conducted fundamental research in both quantum theory and general relativity, laying the groundwork for several subareas of theoretical physics. A generous collaborator and mentor through the years, he had helped launch the careers of many prominent modern theoretical physicists, including Kip Thorne,[21] in addition to Feynman and others.

At the time of my sabbatical leave Wheeler held a joint appointment as professor emeritus at Princeton University and director of the Center for Theoretical Physics at the University of Texas in Austin. I wrote Wheeler in Austin, telling him about my background and asking if he would agree to my spending a year as a Ford Foundation fellow working with him at the center.

After calling my department head, Joseph Budnick, to find out "what sort of person" I was (Budnick, characteristically blunt, said he told Wheeler I had my "head screwed on right"), Wheeler officially invited me to spend my sabbatical as a visiting scholar at the Center for Theoretical Physics.

I sent in my application for a Ford Foundation Senior Postdoctoral Fellowship, explaining I had been invited to work with John Wheeler and his Center for Theoretical Physics. I was awarded the fellowship for the academic year 1982-1983, no doubt largely because of Wheeler's esteemed reputation.

At the start of our journey to Texas in the summer of 1982, Wheeler and his wife, Janette, graciously invited Dorothy and me to visit them at their High Island retreat near Bar Harbor, Maine. We spent a lot of time trying to find the Wheeler home, as it was located in an isolated, rural area. When we finally pulled up next to the house, a jolly-looking man bounded up to the car to greet us. It was the legendary physicist himself. Although I was nervous at this first meeting, Wheeler's engaging smile and friendly manner quickly put me at ease. Comparing notes later, Dorothy and I agreed that he reminded us of the British actor Edmund Gwenn, who played Kris Kringle in *Miracle on 34th Street*.

When we arrived, Janette was not home and our host invited Dorothy to join us in the den for our first discussion. Wheeler's opening comments, it turned out, were not about physics, but about God's perspective on the universe. Soon Janette arrived and she was every bit as friendly as her husband. She invited Dorothy to go with her for a walk. Dorothy admitted to me later that she had been disappointed to leave the discussion because she found Wheeler's comments "so interesting and nontechnical." Indeed, I would come to learn that Wheeler, at home with friends or at the university with scientific colleagues, loved asking big questions about the universe and why it is constructed as it is, questions such as "Why the quantum?" Typical Wheeler follow-up questions would be: "Why does quantum theory underlie reality?" and "Could reality be based on some other fundamental principle?" Other well-known quotes of Wheeler's: "Without an observer, there are no laws of physics," and "How does something arise from nothing?" Such statements and questions are as much a part of philosophy as they are of physics, which gives an accurate measure of the man who asks them with such exuberance.

After a pleasant weekend Dorothy and I embarked on a cross-country trip to the Rockies, then down into Texas.

Austin, home of the University of Texas, is located near the center of the Lone Star State in an area known as the Hill Country. With its gentle rolling green hills, this region is much different from the more typical flat desert and sagebrush terrain that Texas is known for. The compact campus of the University of Texas had, at that time, a student population of about 45,000.

I arrived at the Center for Theoretical Physics just before the start of the fall semester. Wheeler was still in Maine, but his administrative assistant helped me get settled by providing me with office space with another postdoctoral visitor. My office mate was working with Steven Weinberg, who had won his Nobel Prize three years earlier and only recently accepted an appointment as a faculty member in the physics department at the University of Texas.

The son of a court stenographer, Weinberg, born in New York City in 1933, had won the 1979 Nobel Prize in Physics (shared with Sheldon Glashow and Abdus Salam) for contributions to the theory of the unified weak and electromagnetic interaction between elementary particles. While I had been at United Technologies, my boss, Jerry Peterson, had sent Weinberg some of my independent research on unifying gravity and the weak nuclear interaction. I had been thrilled with Weinberg's reply, however guarded it might seem in retrospect, that although there were "some problems," my ideas were "interesting."

Whatever fantasy I harbored now about developing any kind of friendly relationship with Weinberg would soon be put to rest. The first time I crossed his path in Austin, Weinberg, a tall, heavy-set man with reddish blond hair, whisked past me in my shared office to remind his visiting professor about a party Weinberg and his wife were having at their home. Weinberg barely looked at me as he left. I was a little hurt but passed it off.

Actually, it was not the first time Weinberg snubbed me. I had attended a particle physics conference at the University of Pennsylvania about a year after I started at UConn. When I saw Weinberg in the hallway, I went up to him and thanked him for the kind comments he had made about my work in his letter to Jerry Peterson. He looked at me, said nothing, turned his back and walked away. I was confused and hurt and didn't know what to think.

Our next encounter in Austin was at the faculty club. Weinberg entered and raised his hand my way in greeting. I happily returned the wave. He made a point to gesture that he was waving hello to the person next to me. Perhaps I was being too sensitive, but these repeated encounters bothered me. Had I come this far only to be treated with disrespect by a physicist whose work I greatly admired? I mentioned the disturbing incidents to Wheeler's administrative assistant, but asked that she not bother Wheeler about them.

One day, after Wheeler was back on campus, he and I entered an elevator together. (As casually dressed as Wheeler was in Maine, he always wore a business suit at the university, and looked more like a successful bank president than physicist.) The elevator had another occupant: Weinberg.

"Steve," Wheeler said, "I'd like to introduce you to my colleague, Ron Mallett."

Weinberg turned to me, and tentatively shook my hand as though this was the first time he had ever laid eyes on me.

Later, Wheeler's administrative assistant admitted that she had spoken to Wheeler about Weinberg's unfriendly attitude toward me. In this and other ways, Wheeler consistently treated me with kindness and respect, proving that the biggest minds in science can have equally large hearts. Unfortunately, I cannot say that the brilliant Steven Weinberg is one of them. Somehow, Weinberg and I were able to avoid each other for the remainder of my stay.

Dorothy and I moved into a comfortable home in the Westlake Hills area of Austin, renting from a faculty member who was on sabbatical in Italy. Since I would be busy on campus, Dorothy got a job working for a temporary agency and ended up doing clerical and word processing duties for the state human resources department. The community of Austin is very cosmopolitan, and we felt accepted as an interracial couple wherever we went. However, the rest of Texas was another matter. As soon as we left the environs of Austin for the vast flatlands, we felt people constantly staring at us. This was especially so in small towns whose names I have long since forgotten but that sounded a lot like Tumbleweed and Jack Rabbit. In one such place, we walked out of a store and a local looked at me and Dorothy, then showed us an ugly smile. He cocked his hand, pointed it at my chest like it was a gun, and pulled the trigger. This sort of thing did not let me forget for long that I was in the South once again.

My work at the Center for Theoretical Physics was every bit as exciting as I had hoped. It was here that I was to first encounter in any detail the work of Stephen Hawking, which would lead me into a whole new direction of research that would eventually have major implications on my time travel research.

Hawking (born in 1942) had rapidly emerged as one of the most important figures of the late twentieth century in general relativity. He is amazing not only for his brilliant work but also for the circumstances under which his contributions are made. In 1963, while working on his Ph.D. at Cambridge University in England, Hawking was diagnosed with a motor neuron disease called amyotrophic lateral sclerosis (ALS), more commonly known as Lou Gehrig's Disease after the famous Yankee first baseman who died from it. This disease leads to progressive immobility and eventual death. Despite this, Hawking has carried on a wide-ranging research program that has led to fundamental insights from the creation of the universe to the behavior of black holes. One of his most important contributions has been the merger of quantum mechanics with the theory of black holes.

In classical or normal black hole theory, after the formation of the black hole from stellar collapse, matter inside is forever trapped. An example of a normal black hole is known as Cygnus X-1. A black hole alone in space cannot be seen. However, many black holes appear as companions in a double or binary star system, in which two stars orbit each other. One such binary star system was first observed in the constellation Cygnus and designated as Cygnus X-1. This system is about 10,000 light-years from earth. The two stars that make up the system are an optical star that can be seen from the earth, along with an invisible companion in orbit around the optical star. The invisible companion has a mass that is about eight times that of the sun. The gravitational field of the invisible companion is so strong that gases from the optical star are torn from the surface of the star and pulled into the invisible companion. As the invisible companion swallows up the gases from the optical star, they heat up and create a glow around the invisible star that appears as a halo. The center of the halo, where no light can emerge, marks the location of the black hole. Eventually, the black hole will draw the entire optical star (visible) into itself and all that will remain is a single large black hole. The matter and light of the original optical star can never emerge from the black hole. At least this is the case for a *normal* black hole. Then, in 1974, Hawking dropped a bombshell.

Hawking pointed out that if one takes into account that the matter inside the black hole obeys the laws of quantum mechanics, then the game changes. One of the ways of understanding what happens goes back to my earliest fascination with quantum mechanics when I first encountered Schrodinger's equation in the air force. Schrodinger's equation describes the wave properties of matter with a quantity called the wave function and is denoted by the Greek letter psi, ψ. This wave function is a central property of quantum theory and leads to some strange behavior for matter.

While the wave function doesn't reveal the exact location of a particle, it does tell us where it is probably located. This is connected with Heisenberg's uncertainty principle, which does not allow both the location and motion of a particle to be known.

One of the most unusual phenomena associated with quantum mechanics is known as tunneling. In the everyday world, if you throw a ball against a wall the ball bounces back. In the world of quantum mechanics, as a particle approaches the wall the wave that represents a particle has a probability of being on the other side of the wall. This means that as a particle approaches the wall it can tunnel through the wall and appear on the other side. There are practical electronic devices called tunnel diodes that make use of this property.

What Hawking did was to apply this property of tunneling to the matter on the inside of a black hole. He was able to show that if the black hole was small enough, then the black hole barrier would be thin enough for matter to tunnel through and appear outside the black hole. In other words, from the standpoint of quantum mechanics, a black hole can leak matter. As matter leaks, the black hole becomes smaller. Eventually, this leads to the black hole evaporating away. The process of matter and energy emerging from the evaporating black hole is called Hawking radiation. The idea of an evaporating black hole astonished the physics community, and completely fascinated me.

I learned a lot during my stay at Wheeler's Center for Theoretical Physics, mainly in the area of acquiring a deeper understanding of black holes. I hoped that by continuing my inquiry into this strange phenomenon that seemed to warp time I might be led, directly or indirectly, in a fruitful direction regarding the study of time travel into the past. I was excited to get back to Connecticut and continue to explore the strange world of evaporating black holes. In the spring of 1983, Dorothy and I packed our bags and headed for home.

In the early eighties, developments were occurring in the frontiers of cosmology and elementary particle physics that were to have a major impact on my research. Cosmologists have been faced with a problem that the Big Bang theory alone, which states that the universe emerged from enormously dense and hot state nearly 14 billion years ago, did not seem to explain. Why is the universe, on the whole, so smooth? When we look out in space, we look back in time and the further we look out in space the closer we come to the way the universe looked in the beginning. The universe in the beginning looks pretty uniform. Why is that so? It was the work of an MIT physicist named Alan Guth that finally shed some light on this problem. Guth suggested that shortly after the Big Bang the universe went through an extremely rapid expansion, which he called inflation. This inflation essentially erased all the nonuniformity that might otherwise be observed in the early universe. The inflationary universe brought a new understanding to how the universe evolved.

In my research, I began to see that Guth's inflationary universe might affect Hawking's evaporating black holes. Hawking had indicated that black hole evaporation would be noticed if black holes were sufficiently small. Normal black holes are so large that the thickness of the black hole barrier would severely decrease the probability of particles tunneling through the barrier. On the other hand, Hawking suggested that small black holes could be created from the compression of matter at the beginning of the universe. These small early universe black holes are called primordial black holes. Typically they could weigh about a billion tons (around the mass of a small mountain) and be about the size of the nucleus of an atom.

Normal black holes are represented by a solution of Einstein's gravitational field equations named after the German astronomer Karl Schwarzschild.[22] The Schwarzschild solution gives a good description of ordinary black holes. However, the Schwarzschild solution does not change with time. On the other hand, Hawking's evaporating black holes change rapidly with time.

While I was at the University of Texas, I learned that an Indian physicist named Vaidya had found a solution of Einstein's gravitational field equations that could represent a rapidly evaporating Hawking black hole. This solution is called the Vaidya solution. As it turns out, the inflationary stage of Guth's theory is described by de Sitter's expanding solution of Einstein's gravitational equations. It occurred to me that since Hawking's primordial black holes are created in the early universe, the rapid expansion of Guth's inflationary universe could affect the behavior of an evaporating black hole. I knew this would require finding a new solution of Einstein's gravitational equations that would describe both Hawking's black holes and Guth's inflationary universe.

In order to find a new solution, I looked carefully at the physics behind Guth's inflationary universe. In the inflationary model of the universe, a term appears which is called the "cosmological constant." This is the same term that I had encountered when I had studied the de Sitter expanding universe as part of my Ph.D. thesis at Penn State. I thought that if I could somehow combine the cosmological constant into the equation of the Vaidya solution, which describes the Hawking evaporating black holes, then I would have the new solution I sought. After weeks of working with the equations, I succeeded in finding the new solution, which I set out in a paper entitled "Radiating Vaidya Metric Imbedded de Sitter Space," published in 1985 in the *Physical Review.*

That work led me to more research, resulting in a paper on the "Evolution of Evaporating Black Holes in an Inflationary Universe," published the following year. In this paper, I show that the inflation of the universe decreases the radiation from a Hawking black hole. In other words, due to inflation, an evaporating black hole would actually live longer.

This was a highly productive period for me in terms of conducting original research, and I give due credit for the synergism and resultant burst of creativity to that invaluable year I spent in the orbit of one John A. Wheeler. In 1987, I was promoted to full professor.

In the mid-1980s, I had the good fortune to come in contact with a brilliant graduate student named Kristine Larsen. Kris had recently graduated magna cum laude with a bachelor's degree in physics from Central Connecticut State University. After attending a talk I had given at Wesleyan University, she showed up at my office and said she had decided she wanted to do her Ph.D. thesis at UConn under my supervision. She did not yet have a topic in mind, and we discussed a number of possibilities. Nothing immediately jelled. Then, she mentioned that she was interested in the work of Stephen Hawking, and planned to attend an upcoming conference in Chicago where he was scheduled to speak. As it turned out, I was planning to attend the same conference. We decided to compare notes about a possible thesis topic for her after hearing Hawking.

The conference was an astrophysical symposium on relativistic astrophysics. Kris and I met there, and went to Hawking's talk together. He was completely wheelchair-bound at this point and could only communicate via a computer-generated voice with a menu chosen on a computer console attached to his wheelchair. That apparatus was soon forgotten as one listened to the elegant science. As Hawking has said, "I would rather be known as a scientist who happens to be disabled than as a disabled scientist." Beyond a doubt, Stephen Hawking is first and foremost a scientist.

After a banquet on the final night of the conference, I returned to my room. Tired from several long days, I went to bed early. About an hour later the phone jarred me awake. It was Kris.

"Ron, get down to room 520 now!"

"Why?" I asked.

"It's Hawking's room, and we're having a party."

Needless to say, it didn't take long for me to dress and run to the party. In fact, Kris, in retelling the story through the years, would claim that I must have exceeded the speed of light in getting to Hawking's room.[23]

When I arrived, there was already a crowd of people drinking wine, beer, and mixed drinks. Everyone, including Hawking, seemed to be enjoying themselves greatly. Hawking was friendly and expressive. His large head rolled to one side and his eyes, which were magnified by his glasses, seemed to twinkle with a certain mischievousness. Having read that he enjoyed music, I asked, "Who is your favorite composer?" He communicated via his computer console by highlighting words and phrases from a menu with a cursor. After tapping on his console, his answer came forth: "Wagner." (I learned later that shortly after he was diagnosed with ALS, Hawking became depressed and spent hours listening to Wagner's *Götterdämmerung* [*Twilight of the Gods*]. It seems Hawkins and I share another interest besides music. I am told he has on the wall of his office pictures of Marilyn Monroe.)

Not long after, Kris and I, along with some other UConn colleagues, attended a physics seminar at Yale University where we heard a talk given by a collaborator of Alan Guth's. The lecture was about Guth's recent work on "child universes." Guth had theorized that on the other side of black holes a whole new universe may be in the process of creation. Since the black hole was located in our universe, the new universe was dubbed by Guth to be a child universe. As I saw it, the problem was that since the child universe was created on the other side of a black hole and we can't see into a black hole how do we know if a child universe is really being created?

At that point, I turned to Kris and said, "This could be your thesis."

Deciding to go directly to Guth and ask what he thought about the idea, I spoke with him at a conference held at Harvard. I told him that I was considering having my graduate student work on the problem of theoretically determining whether a child universe could be observationally detected. This would be done by looking at child universes that were formed on the other side of a Hawking radiating black hole. The process of creation of the child universe should affect the radiation from the black hole that we see in our universe.

Guth, friendly and outgoing, listened intently, and said he thought it was a good idea. Furthermore, he said that as far as he knew, no one had worked on the idea before. He then generously offered that, if we wished, my student and I could visit him from time to time to discuss the progress of the work.

When I told Kris about Guth's response, she was thrilled. She started her research while working in the physics department as a teaching assistant, then received a paid graduate assistantship which allowed her to devote herself full-time to her research without any classroom duties.

On a regular basis, Kris and I made a pilgrimage to MIT to discuss her latest results with Guth. Although he could be intense when discussing questions of physics, Guth had an easy smile and was enjoyable to spend time with. He seemed to have a fondness for Coca-Cola, as his office was lined with stacks of empty cans.

In a brilliant tour de force, Kris was able—using the solution of Einstein's gravitational equations that I developed earlier that studied evaporating black holes in the early universe—to show that there was indeed a very characteristic change in the radiation rate of an evaporating black hole during the process of child universe creation. Guth liked her results, which were published in *Physical Review* in 1991 in a paper entitled "Radiation and False Bubble Dynamics." Kris received her Ph.D. in 1990. Even before she finished her doctorate, she was offered a position on the physics faculty of her alma mater, Central Connecticut State University.

I was proud of Kris's thesis work and of the fact that it was based on the new solution of Einstein's gravitational field equations that I had found in my study of Hawking's radiating black holes in an expanding universe. This solution had required that I develop an expertise in solving Einstein's equations. Like a stepping stone, it was this knowledge that would prove to be crucial to my eventual breakthrough in developing a new theory of time travel.

Before reaching that pinnacle of my life's work, however, I would have to find my way through some black holes of my own.

Eleven

Back from the Brink

⧗

Since my father's death I have been a loner. That life-altering event not only changed my personality, it also triggered my first lasting depression. At the time, I had found a degree of solace only in being alone, reading science fiction, and daydreaming incessantly.

In high school I suffered another depression, although it might have been a long, continuous one since childhood. After experiencing some exhilaration at being away from home and in the air force, I sunk into a depressive state while in the service. Feeling antisocial, I tried to cut myself off from the rest of the world, which was one reason I volunteered for the graveyard shift in the computer control room. Once again, I found escape in books, and also in academics. It wasn't until I entered Penn State, and met Marjorie, who came to represent the ideal of beauty and brains, that I began to see the possibility of having a fuller life.

Now, after a rewarding decade during which I became a full professor, conducted original research (much of it published) that increased my knowledge of science relevant to time travel, and enjoyed a stable marriage, I harbored growing feelings of discontent. I was filled with questions I could not answer. *Would I ever accomplish my life's goal? Was I losing my way?* One lesson I had taken from the sudden loss of my father was that some opportunities exist for only a finite time and one has to take advantage of them whenever they present themselves or possibly lose them forever. *Had I led a fantasy-fueled life all these years? Had I wasted my time studying so hard for something that I would never accomplish?* Gradually, I slid into a full-on depression.

At the university, I hid my dark moods as best I could. Putting one foot in front of the other, I taught classes, prepared course work, attended meetings, interacted with colleagues, and guided students—all the while feeling wretched. At home, I was morose and melancholy. Reminiscent of my earlier depressions, all I wanted was to be left alone. Without saying a word to Dorothy, I would get up from my chair in the living room, go into the bedroom, shut the door, turn off the lights, and lie in the dark staring at the ceiling. I feared the sum of my life had fallen drastically short, not only personally but professionally. Certainly, my dream to accomplish breakthrough time-travel research and build a working machine had not come to fruition, and at times I felt about as far away from that goal as I had been at age twelve, when I built my first rudimentary time machine in the basement of our home with my father's tools using illustrations from a comic book as my blueprint.

During these low periods, the only thing that helped were positive memories of my childhood, always set in the years before my father died. Over and over, I reran scenes such as Dad showing me the inside of a TV set as he patiently explained how the yoke supports the picture tube. There was much he didn't have a chance to show me, and many questions that I didn't get to ask him.

My depression took its toll on our marriage. Dorothy, a gentle woman who had always represented security to me, tried to help.

When she saw me grieving for my father, she would say out of sympathy and exasperation, "Can't you let the spirit of Boyd Mallett rest in peace?" But I couldn't. Even though I had always felt a sense of safety with Dorothy, there was little she could now say or do that had any impact on me. Hitting rock bottom emotionally, I yearned to escape and be alone, as I had in the past. We never had arguments; I simply withdrew from Dorothy and our marriage, which deeply hurt her. We finally agreed that a separation was in order and I should get my own apartment. In hindsight, it would have been wise for me to see a therapist. But at the time, leaving seemed the only option. Before the fall semester started in 1989 I rented a small apartment in Manchester and began living alone for the first time in more than twenty years.

Dorothy and I checked in with each other regularly, and met periodically. She longed for more stability in her life than I was able to provide at that time. It became apparent we were growing apart. We divorced in January, 1991. Missing her family, she moved back to Pennsylvania a few months later. Not long after the move, she met a successful businessman; they eventually married, happily so.

During my years of living alone I took up the piano. Listening to classical music had always been a source of comfort for me. At my first piano lesson there was something about the form and flow of musical notes on a score that reminded me of mathematical equations. My piano teacher, Judy Rodwell, predicted that, because of the mathematical basis for the laws of musical harmony, my background would serve me well in learning to play. I soon developed a passion for the keyboard, and would spend many hours each week practicing. I eventually experienced the satisfaction of mastering Chopin's *Prelude in E Minor*. This somber piece spoke to me of longing and loss. Somehow, it gave my feelings an avenue of expression. Playing the piano became cathartic. Coming home from work, I would sit at the keyboard and play nonstop, sometimes forgetting to eat or even turn on the lights. Slowly, thanks in large measure to music, the hole in my soul began to close back up. In small doses I started enjoying life again, seeing old friends, making new ones.

The day came, after four years of bachelorhood, that I was ready to venture forth. Rather systematic in everything I do, I enrolled in a course called Fifty Ways to Meet a Lover. Thereafter, I took out a singles ad with the title "Adventurous Astrophysicist Seeks Interesting Terrestrial Female." Nevertheless, I met Deborah McDonald, an intelligent and sophisticated woman with two children, the old-fashioned way: at a church social. Although I had always wanted my own offspring, that was not to be. The best alternative was finding a loving, supportive partner who arrived in my life with an instant family. Deborah (then working as a state lobbyist for a nonprofit group) and I married the following year (1993), and I happily became stepfather to Sarah, sixteen, and Andrew, thirteen.

There was a wonderful closeness between Deborah and her children that she willingly let me become part of. One of our most memorable summer vacations as a family was a visit to Einstein's home at 112 Mercer Street in Princeton, New Jersey. Sarah and Andy knew that my work at the university was somehow con-nected to Einstein's ideas, and they were excited and proud. Even though the house wasn't open to visitors, the four of us stood together on the porch and had our picture taken. Sarah and Andy were the best kids a stepfather could want. I am proud that they both became teachers.

One summer evening in 1996, as Deborah and I were finishing a pizza while watching a movie at home, I felt an excruciating pressure on my chest. I tried to relax, hoping it would go away. When it didn't, Deborah drove me to Manchester Memorial. At the hospital, I had the good fortune to come under the care of an astute cardiologist, Dr. Parveen Khanna, who recognized from the electrocardiogram that there was a major problem with my heart. He ordered an angiogram, and determined that I had what is referred to as a "widow maker," a blockage of the major artery going into the heart. In my case, it was 95 percent blocked. Angioplasty surgery was done and a stent inserted in the artery to correct the problem. Although physically I soon felt as good as new, I struggled with new feelings of mortality, common among heart patients.

My mother—a widow for the second time and still residing in the same house in Altoona—had provided me with added details about my father's health. Unbeknownst to me when he was alive, my father had a heart condition for which he took medication. She also confirmed that he had at times been sad and depressed over the possibility of an early death like his own father. As I processed that information, I could become angry at him for abusing his weakened heart with a two-pack-a-day habit. Didn't he *know* about the link between heart disease and smoking? Or had that information not yet been widely disseminated in the 1950s? And yet I had never smoked cigarettes and I was having heart problems at fifty years of age. Was I, too, doomed to die early? Aggravating my feelings of mortality was the deep disappointment I felt at my continued lack of progress in time travel research. Irrespective of how long I lived, I knew I did not have all the time in the world to realize my dream.

Providence eventually intervened, providing a crossing of paths with one Fred Adams, professor of physics at the University of Michigan, at a 1998 conference held at the University of Kentucky. We hit it off immediately. Adams and I met for beers one evening in the hotel bar. A world-renowned theoretical astrophysicist who had studied star formations and cosmology, Adams tackled big questions like the origins of existence of the universe. We talked about our work and careers. Adams is an easygoing guy with a friendly demeanor and quick smile, and soon I was reciting the whole story of how I became interested in time travel and my very personal motivation for becoming a physicist. I acknowledged how I had been keeping most of this close to my vest for years and had only told a handful of intimates, afraid it would amount to professional suicide in academia. He said there was a lot of work going on in time travel, and he encouraged me to "check the literature." He thought I should get back to pursuing my longtime goal to design and build a working time machine.

It's strange how it happened—at a low point having my inspiration rekindled by such a chance meeting. I left Kentucky determined to make a new and concerted effort to review all the current literature and research having to do with time travel. Over the course of the next year and a half, I was frankly surprised at how much work was being done in the field by others, including some of the biggest names in science.

In searching the literature I came across an article published in 1974 in the *Physical Review* entitled "Rotating Cylinders and the Possibility of Global Causality Violation," by Frank J. Tipler (presently at Tulane University). By analyzing a solution of Einstein's gravitational field equations published in 1937 by University of Edinburgh physicist W. J. van Stockum, Tipler found that the region outside an infinitely long, rapidly rotating massive cylinder contained "closed time-like lines." As with the situation involving Gödel's rotating universe and Kerr rotating black holes, the presence of closed time-like lines indicated the possibility of traveling in time to the past.

Further study led me to a 1988 paper by Caltech physicist Kip Thorne and his collaborators, M. Morris and U. Yurtsever, in *Physical Review Letters*. The manuscript was entitled "Wormholes, Time Machines, and the Weak Energy Condition." Wormholes have a long history that goes back to a 1935 paper by Einstein and Nathan Rosen, in which they found a solution of the gravitational field equations that allowed for the possibility of a tunnel between our universe and another universe. The tunnel was called an Einstein-Rosen bridge. In 1955, John A. Wheeler had suggested a tunnel not between two different universes but between two different parts of our universe, and had coined the term "wormhole" to describe this tunnel.

A wormhole can be thought of as a shortcut between different parts of the universe, essentially through space and time. A wormhole has at least two mouths which are connected to a single throat. If the wormhole is traversable, matter can travel from one mouth to the other by passing through the throat.

A demonstration of a wormhole can be accomplished by using a small rubber ball. First, mark one point on the ball as point A. Then, mark the opposite end of the ball as point B. Now drill a hole directly through the ball from point A to point B. There are now two ways to get from A to B: the long way around going along the surface of the ball or a shortcut directly through the ball. The tunnel drilled directly between A and B is like a wormhole. In the same way, a wormhole is a direct tunnel between different points that does not lie along the surface of space.

In theory, a wormhole can be used as a rapid transit route from one part of space to another that would take too long by going along the surface of ordinary space. This was the basis for mankind's first contact with an extraterrestrial civilization in Carl Sagan's science-fiction novel and movie, *Contact*. This mode of travel was suggested to Sagan by his friend Kip Thorne as scientifically plausible.

In their 1988 paper, Thorne, a former student of Wheeler's, and his collaborators showed how a wormhole could be used as a time machine, one built by nature rather than man. This could be accomplished by accelerating one end of the wormhole relative to the other; relativistic time-dilation (the time on a clock moving near the speed of light is slower than that of a stationary clock) would result in less time having passed for the accelerated wormhole mouth compared to the stationary one. This means that an object which entered the accelerated wormhole mouth could exit the stationary end at a point in time prior to its entry. Such a path through a wormhole is called a "closed time-like curve." A wormhole with this property could be thought of as a hole in time.

Continuing my literature search, I turned up yet another mode of time travel into the past that had its roots in the origin of our universe. In 1991, Princeton physicist J. R. Gott published a *Physical Review Letters* article entitled "Closed Time-like Curves Produced by a Pair of Moving Cosmic Strings: Exact Solutions." By then, everyone in the physics community knew that a "closed time-like curve" was code for *time travel into the past*. Essentially, Gott's paper was about a cosmic-string time machine.

Cosmic strings can be thought of as fault lines in the structure of the universe left over after the Big Bang. These strings are infinitely long massive threads that could exist anywhere in the universe. Gott considered two infinitely long parallel cosmic strings moving toward each other. As the strings pass each other they create a closed loop in time. It is along this loop that time travel into the past can take place.

The papers of Thorne, Gott, and Tipler had a liberating influence on me. Here were well-known, highly regarded members of the physics community publishing papers about the methods and means of time machines in respected scientific journals. I felt inspired to renew my efforts and, for the first time in my career, work openly on the problem of time travel. However, due to my teaching and other academic responsibilities, I lacked the time to completely engage my energies to this task. That was about to change unexpectedly.

One gratifying characteristic of the UConn physics department is its collegiality. I've always felt comfortable among the faculty, staff, and students. Dan McLaughin was a former graduate student in the department who was now in his forties and teaching at the University of Hartford. I liked Dan. He was a big man with a ready smile and a large bushy moustache like men wore in the late nine-teenth century. Having recently been married, Dan was clearly enjoying life. Just before Christmas 1998 I ran into him outside my office. We talked for a while, then wished each other happy holidays.

Two weeks later, I received a phone call from an administrative assistant in the physics department. Tearfully, she told me that Dan had died of a sudden heart attack. I couldn't believe it. He had looked so alive and well two weeks earlier. Now he was dead and gone?

Emotions connected with my father's sudden death were stirred up inside me. Panic set in. My own heart condition made me think that I, too, might suffer a fatal heart attack. Sure enough, I soon began to experience severe chest pains. After consulting with my cardiologist, and with the approval of my department head, I was granted a six-month medical leave of absence.

At first I spent my time at home in a familiar state of depres-sion. Not even the piano seemed to help. Constantly concerned about my heart giving out, I was consumed with thoughts of death. Then one day, still in my robe and sleepers, I wandered into my study. I began to pore over what I had learned about various time travel theories. Over the course of several afternoons, I became engaged with my favorite subject again. I started doing calculations on pads of paper that I kept around the house, including next to my bed. The stress that seemingly had control of my body released its grip and started to fall away.

Now free of my normal academic responsibilities for months, I found myself with all the time I needed to focus my attention on the topic of time travel to the past. After a couple weeks of exten-sive reading, I was left curious, and wondering: *Was there another approach, besides the methods I had studied, that might lead to a new concept of a time machine?*

I went back to basics. In order to know where a time machine is traveling, it is necessary to know the direction of time. In relativity, light is the basis for separating the past from the future. When you strike a match, the light spreads out in all directions. Even though light moves at a fast clip of 186,000 miles per second, it still takes time for the light from the match to spread out like the outward moving ripples in a pond. The future direction of time is determined by how far light has moved from the match. The spreading of light from the match can be represented as an expanding circle increasing with time. As time goes on, the circle gets larger and larger. Lines drawn along the edge of the expanding circles form a cone. The cone would look just like an ice cream cone. In fact, if you took a piece of string and threaded it through the tip of the ice cream cone, the forward direction of time would be the direction of the string from the tip of the ice cream cone to the open mouth of the cone. The spreading cone of light is called a light cone. The widening of the light cone is the direction of the future.

In special relativity, space and time are flat. This means that the time line that goes from the past to the present to the future is a straight line that is threaded through the light cone. The apex of the light cone represents the present moment (for example, the moment you strike a match). Any point along the line after the apex of the cone as it widens represents a future event, whereas any point along the line before the apex of the cone represents a past event. As a person moves into the future, the apex of their light cone (the present moment) moves forward. It's like moving an ice cream cone along a straight string that has been threaded through the tip of the cone. It would be vital for me to keep the direction of time in mind as I worked toward developing a theory for building a man-made time machine.

During this period, I also studied Caltech physicist Richard C. Tolman's classic 1934 textbook "Relativity, Thermodynamics, and Cosmology." In Newton's law of gravity, only matter can produce a gravitational field. For example, the earth produces a gravitational field that keeps you anchored to the ground. In his text, Tolman pointed out the surprising result that in Einstein's general theory of relativity not only matter but also light can be a source of gravity. Tolman considered the gravitational field of what he called a straight "thin pencil of light." In Tolman's era,[24] a thin pencil of light could not be produced. Light spreads out from all natural sources and even the light from a flashlight beam fans out from the opening. As a laser produces an extremely narrow beam of light, I wondered if this meant that modern laser technology could be used to produce Tolman's thin pencil of light.

As I pondered this, I thought about the common denominator among Gödel's rotating universe, Kerr rotating black holes, van Stockum-Tipler rotating massive cylinders, Thorne's wormholes, and Gott's cosmic strings. With the exception of the wormhole, a common factor in them seemed to be the relative rotation of matter. It was then that I had a Eureka moment.

I knew from my time working with lasers at United Technologies that there was a device called a ring laser that could produce an intense and continuously circulating narrow beam of light. It occurred to me that this continuously circulating light beam might produce gravitational effects similar to that of rotating matter.

My earlier studies of Kerr rotating black holes indicated that the rotating matter of the black hole resulted in a dragging of space around the black hole, rather like a rotating apple in a vat of molasses. This dragging of space is called frame dragging. In addition, the rapidly rotating black hole also leads to the formation of closed loops in time.

Studying the mathematical structure of the rotating black hole in more detail, I noticed that parts of the equation for the rotating black hole that led to frame dragging also appeared in the equation for a black hole that led to closed time-like lines. This suggested that frame dragging was related to closed loops in time.

That was important. If I could show that the gravitational field of a circulating light beam in a ring laser could produce frame dragging, this could also imply that the circulating beam of light could lead to closed time-like loops.

Calculating the gravitational field of the circulating light beam would require all the skills I had developed in solving Einstein's gravitational field equations in my earlier work on black holes and cosmology. I began the process of finding a solution for the gravitational field of light in a ring laser.

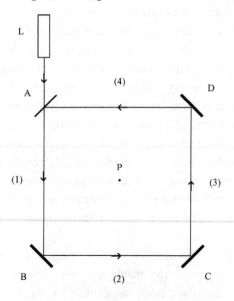

A sketch of the ring laser geometry.

By now, I had become totally immersed in the problem, and was no longer conscious of my physical state. This is usually what happens to me when I am trying to solve a problem. I ate and slept erratically, and often worked into the wee hours of the morning. I don't remember dreaming about the equations—as I had done previously at such times—but I do recall lying awake in bed trying to figure out how to calculate a particular term in the equations.

To do the calculations I chose a standard form of the ring laser. This configuration consisted of a square with a mirror at each corner of the square. One of the mirrors is semi-transparent while the other three mirrors are reflecting. A laser beam enters the square through the semi-transparent mirror. The laser beam is then reflected by each corner of the square and returns to the original mirror where the beam once again is reflected to the other mirrors. In this way, a circulating beam of light is created around the square by the continuous reflection of the light from each corner of the square.

Day after day, I scribbled furiously on endless pads of paper in my den, working at solving the gravitational field equations for the area inside the square of the laser ring. To simplify the calculations, I studied the situation when the gravitational field is weakest. This involved using a standard approximation scheme for Einstein's equations in a weak gravitational field. Sometimes, physicists use mathematical techniques called "approximations" to simplify equations. If the effects they are looking for are small enough they can skip additional terms in an equation. For example, consider an equation with two numbers: .1 and .0001. Because the second number in the equation is so much smaller than the first number, it can be disregarded. That is an approximation.

In Einstein's gravitational field equations, if the gravitational field is sufficiently weak it is possible to simplify the calculations by skipping everything except the largest term in the equation. So, I first applied the standard approximation for a weak gravitational field to Einstein's equations, reducing the number of terms in the equations. Then, to calculate the gravitational field of the ring laser, I separated the square that formed the ring into four segments. Using a mathematical summation technique called integration, I found the gravitational field due to the energy and flow of the laser light for each segment. Putting the individual segments back together, I could then find the total gravitational field for any point inside the ring laser.

To judge the effect of the gravitational field, I considered what would happen if I placed an electrically neutral mass spinning inside the ring. This meant calculating equations for the spinning particle in the gravitational field that I had worked out for the ring laser. The calculations were tedious, and I made some mistakes along the way, which meant redoing the calculation. As I worked out my results, I would often listen to music from Richard Wagner's operatic masterpiece, *The Ring of the Nibelung*. The fact that Wagner's heroic score dealt with a mystical ring gave me added inspiration.

After several weeks of intensive work I found that the light beam in a ring laser produced a gravitational field that looked— mathematically, at least—like a vortex or twisting of space not unlike the swirling water going down a bathtub drain. This twisting of space resulted in the dragging around of a subatomic particle, such as a neutron, when placed at the center of the laser ring. My calculations showed that the gravitational field of a circulating light beam *could,* in fact, produce one of the important effects associated with the gravity of rotating matter: frame dragging.

I rechecked my calculations, finding no mistakes.

My solutions proved something new and revolutionary. While frame dragging due to the gravitational field of a rotating massive body (such as the earth) had been predicted as early as 1918, no one before had shown gravitational frame dragging due to a circulating light beam.

The presence of frame dragging was significant since it was linked in black holes to the formation of closed time-like loops. It was logical to expect that frame dragging by a circulating light beam could also be linked to closed loops in time. I knew that was the direction I would next take my work.

Eager to share my new scientific predictions with my peers, I wrote up my results and submitted them to *Physics Letters A*. This entailed my having to compress nearly a hundred pages of calculations into a four-page paper for publication. In scientific papers, the author shows only the starting equations, a general outline of the calculations, and the final results. The details of all the calculations are never shown, as they would take up too much space.

In April 2000 (I had returned to teaching the previous fall semester and was no longer having chest pains or experiencing any other physical ailments) I received word from the science journal that my manuscript had been accepted for publication. I know this meant that an anonymous referee who is an expert in the field had checked the equations and found no mistakes. (If any mistakes had been found, the paper would have gone to a second expert. If the second expert also found mistakes, then the paper would have been rejected for publication.)

On May 8, 2000, my paper "Weak Gravitational Field of the Electromagnetic Radiation and a Ring Laser" appeared in print. Once an article is published, it is widely read by the physics community at large. To date, my fellow physicists have not pointed out errors in my math or science. This work represented my first breakthrough in time travel research. Although there was no mention of "time travel" in the published paper, the research was a necessary stepping stone in my attempt to develop the theory and design for a man-made time machine based on a circulating beam of light.

My lifelong dream seemed to be at the threshold of becoming a reality.

Yet, I was not finished.

Twelve

The Sweetness of Technology

⧗

My department head, William Stwalley, said he thought my new research would be a "novel test" of Einstein's general theory of relativity. Another department colleague, Win Smith, paid the highest compliment one physicist can give another when he described my published work as "elegant." Both Stwalley and Smith, however, wanted to know more about the magnitude of the space-twisting effect; specifically, under what conditions would the effect be large enough to measure? I explained to them that since the equations showed that the effect depended on both the laser power and ring-laser size, I was studying different configurations in order to best maximize the effect.

In February of 2001, I received a call from Fred Adams of the Michigan physics department. He wanted to know about my progress since we had chatted over beers in the Kentucky hotel bar three years earlier. I gave him a brief summary of my research, and told him about my *Physics Letters A* article. He asked if I would give a talk about my work at his physics department's colloquium. Because I felt I owed Adams for the nudge he had given me in the right direction, I accepted without thinking about the ramifications. We talked about titles for my talk.

Conservative by nature, I suggested "The Gravity of Circulating Light."

"You've applied this research to time travel," Adams said. "Let's put that in your title."

I pointed out that I had done only the first part of the work, showing that circulating light leads to the twisting of space, and was now working on whether it would also lead to the twisting of time, "a necessary requirement for time travel to the past. I haven't finished all my calculations yet on the second part."

"That's okay," Adams said. "You can explain that your work is still in progress. Why don't we call it 'The Gravity of Circulating Light: A Possible Route to Time Travel.'" Such a title, he explained, would "jazz up" the topic and "attract more attention."

"Fine," I said. It was, after all, his department's colloquium.

As it turned out, that title attracted more attention than either of us could have possibly anticipated.

The physics department at the University of Michigan, Ann Arbor, is well-known internationally. From 1928 through 1941 it was the home of the famous summer symposia on theoretical physics, which brought to Ann Arbor the likes of P. A. M. Dirac (1902–1984), famous as the creator of the complete theoretical formulation of quantum mechanics; Enrico Fermi (1901–1959), winner of the 1938 Nobel Prize in Physics for his work on induced radioactivity, and noted for the development of the first nuclear reactor (Fermi is widely regarded as the only physicist of the twentieth century who excelled both theoretically and experimentally); Kasimir Fajans (1887–1975), a Polish-born U.S. chemist who did pioneering work on radioactivity and isotopes; and many other scientific notables, including Wolfgang Pauli, Werner Heisenburg, Neils Bohr, and J. Robert Oppenheimer. David Dennison (1900–1976), the discoverer of the spin of the proton, spent his career at Michigan. The world's first bubble chambers—devices for detecting subatomic particles by means of a string of bubbles left in a chamber of liquefied hydrogen—were developed at Michigan by Donald Glaser,[25] for which he was awarded the 1960 Nobel Prize in Physics. One of many notable alumni, Samuel C. C. Ting (born 1936), who earned his Ph.D. in physics at Michigan, won the 1978 Nobel Prize in Physics (along with Burton Richter) for "pioneering work in the discovery of a heavy elementary particle of a new kind."

Due to the renown of the Michigan physics department, events and press releases are tracked by the news media, nationally as well as abroad. One afternoon in February, two months before I was scheduled to appear in Ann Arbor, I received a call from London. Physicist Michael Brooks, an editor at *New Scientist,* the international equivalent of *Scientific American,* was calling to find out more about my scheduled talk at Michigan. After a number of lengthy phone conversations, Brooks decided to do a feature article about my work that would come out after I spoke at Michigan.

In the meantime, I had returned to my research, finding as much time for it as I could fit in with my academic responsibilities. Historically, most breakthroughs in science are made at universities where professors have teaching responsibilities. The exception is the Institute for Advanced Studies at Princeton (Einstein, Godel, et al.), which is designed strictly for research.

Having only partially worked out the gravitational effects of circulating light beams, I turned to the aspect that would lead to time travel via closed loops in time. To work out the latter half of my theory would require looking at strong gravitational fields.

In solving Einstein's gravitational field equations for the weak gravitational field of the ring laser, I had used a standard approximation scheme. To find closed loops in time generated by circulating light, I would not be able to resort to approximations. The complete solution of the problem would require finding what are known as "exact solutions" of the Einstein field equations. Finding exact solutions of Einstein's field equations is a notoriously daunting problem. Hoping to simplify the problem, I considered two beams of light circulating in opposite directions. In this way, I arrived at a tentative solution. I was uneasy with having deviated from my original ring laser model with a single beam of light going in one direction, but decided to see where the model of two light beams moving in opposite directions would take me.[26]

In April (2001) I flew to Ann Arbor, which has a population slightly more than 100,000 people and an additional 40,000 students. I settled into a comfortable hotel near the campus. The next morning, my host, Fred Adams, picked me up and took me to the lecture hall where I was to give my first public talk on my new research. About fifty people were already present. I knew from having attended other colloquiums that they would be predominately graduate students and faculty from the physics department, although they could be from all over the spectrum –solid-state physics, nano technology, atomic and molecular physics, etc. The best presentations at such events approached a topic from the middle ground—not overly technical, and yet more scientific than a popular-level talk.

Adams gave the audience a brief introduction.

As I stood waiting, I composed myself with several deep breaths. I realized that I was about to come out of the closet I had sequestered myself in for years. Once the door was thrown open, there would be no going back.

Since the audience was composed mainly of physicists in other specialties, I began with a brief background of Einstein's special and general theories of relativity. I reminded them that in the special theory, Einstein had showed that time slows down for a moving clock. In addition, the general theory of relativity indicates (in part) a slowing down of time in a gravitational field.[27]

In the general theory, gravity results from the bending of space by a massive body like the sun. The earth and the other planets move around in orbits guided by the curved space produced by the sun. I pointed out that in Einstein's theory, even light, which does not have mass, has energy and that energy can cause space to bend. Since light can bend space, this means that light can also produce a gravitational field. I explained that this gravitational effect of light was the key to my discoveries.

I then described how the circulating light of a ring laser caused a twisting of space. I illustrated this by using one of my favorite analogies of a spoon in a cup of coffee. Imagine, I said, that the coffee in the cup was like empty space, and the spoon is like the circulating light beam. As I stirred the coffee with the spoon you could see how the coffee swirled around. In exactly the same way, the circulating light beam is causing a stirring of empty space. If you placed a sugar cube in the coffee it would get swirled around because the coffee is getting stirred by the spoon. Similarly, if you placed a spinning subatomic particle like a neutron in the empty space surrounded by a circulating light beam, you would see the neutron being twisted around like the sugar cube in the coffee.

Next, I pointed out that in Einstein's general theory of relativity, space and time are connected. If space gets twisted, time gets twisted. I explained that I was working on a solution of Einstein's gravitational field equations, which I hoped would show that two light beams circulating in opposite directions would lead to time being twisted in a loop, and that this in turn would lead to time travel into the past. "However," I said, "this research is not yet complete."

When I finished, the response of the audience was cautiously polite, and I wasn't sure what to make of it. There were only a few questions, and no one had any objections to the physics or the mathematics.

When I returned home, I got a call from Michael Brooks, wanting to know how my lecture went. I gave him a summary and suggested he contact Fred Adams. Shortly thereafter, I received an e-mail from Brooks telling me that *New Scientist* had decided to do a cover story about my time travel work. The May 19, 2001, issue of *New Scientist* hit the newsstands with a cover headline: "Flashback: Presenting the World's First Time Machine." The article began: "Ronald Mallett thinks he has found a practical way to make a time machine. Mallett isn't mad. None of the known laws of physics forbids time travel. . . ." The article went on to give an overview of my research, along with these comments by Fred Adams regarding my talk at Michigan: "The reception was cautious and skeptical. But there were no holes punched in it, either. The solution is probably valid."

When I turned on my computer that evening to check my e-mail, I was shocked to see dozens of messages from people I didn't know—and who had apparently gotten my e-mail address from the UConn Web site—inquiring about my time travel research. It took hours for me to answer them. The next day, there were dozens more e-mails. It was then that I began to realize the deep-seated interest many other people have in time travel.

That same week I heard from Ben Bowie, a documentary film director in England who had read the *New Scientist* article. Bowie, who told me Michael Brooks was a friend and neighbor, said he thought my research would make for a good documentary. He wanted to fly to the U.S. to meet me. We arranged for a rendezvous the following week. Having never dreamed my research would serve as the subject of a documentary, I looked forward to meeting the director. But almost immediately, I was filled with dread.

Michael Brooks did not know that I was black, and the *New Scientist* article did not make reference to my race. I knew that Ben Bowie would also not know, as my phone conversation with him would not have given away my ethnic origin. In fact, I had once been told by an African-American administrator that I did not "sound black" over the phone. My experiences growing up in white America had taught me that no matter how high I climbed, it could never be high enough for some people simply because I am black. I feared that when Bowie saw me, he might find some reason not to continue with the project.

As it turned out, my fears were baseless. Bowie was unhesitatingly friendly and interested in my work, not in the color of my skin. As we talked, it was clear that he thought my ideas were sound, and could lead to the development of a working time machine. It was important to him that my work was based on Einstein's theories of relativity, and Bowie felt that a suitable documentary could be made discussing Einstein's general and special theories of relativity while highlighting my research into time travel. Thus began a fruitful enterprise which resulted in *The World's First Time Machine,* a well-crafted science documentary that first aired in 2003 on The Learning Channel, and at the same time on the BBC in England. Using state-of-the-art animations and real-life sequences, the film covers Einstein's special and general theories of relativity and quantum theory and my own time travel research.

Newspapers and magazines were now calling on a regular basis. After years of keeping mum about my interest in time travel because I feared being labeled "the crazy professor," it was astonishing how public my work had so quickly become. To my amazement, the press coverage extended to that arbitrator of rock and roll, *Rolling Stone,* which noted my work in their August 2001 issue under the headline "Hot Theory: Time Travel."

Somewhere early on in the process of being discovered by the media, I was being challenged by a reporter in an interview. As I recall, I was being compared to Doc Brown, the mad professor in *Back to the Future* who invented the Flux Capacitor to travel back in time. In exasperation, I finally said, "Look, I'm not a nut. This is not Ron Mallett's theory of matter. It's Einstein's theory of relativity. I'm not pulling things out of the known laws of physics." In case there was any doubt, I was basing not only my theories but also my career and reputation on the foundations built by Einstein.

A reporter for the *Wall Street Journal* interviewed me, which resulted in a lengthy article, "Physicists Are Looking at How We Might Take a Trip Through Time." In the article I explained about my long-time interest in time travel, and how I had kept quiet and did not come "out of the time-travel closet" for a long time. The reporter then noted: "The closet is emptying fast. . . ."

And how.

The *Boston Globe* ran an editorial titled "Time and Again," which read in part:

> Time is a straight line for most people—yesterday, today, tomorrow. It's what the watch says, what the guy on the radio says, or what the ticket says. It can seem to race or drag, but these are states of mind rather than permutations of the old linear 24/7. But what if the clock were as malleable as Gumby and time machines as common as airports? What if we could hop in and fly?
>
> Ronald Mallett, professor of physics at the University of Connecticut, feeds this science fiction fantasy, which is why he packed 200 people into a Museum of Science lecture hall on a recent Friday night. Traffic-weary, gravity-weighted, skeptical, and curious, they came to hear from someone bold enough, or crazy enough, to suggest that time travel could happen in this century. . . .

Soon afterward, a Boston TV station contacted me about doing a story. The station also called Alan Guth at MIT to ask him about my work. I received an e-mail from Guth suggesting that we get together to discuss my current research. On a hot and sticky August day I traveled to Boston and met with Guth, a journey that I had taken several times before and always enjoyed.

Guth, still the friendly, intent listener I remembered from my previous visits, immediately grasped the basic idea of my ring laser-gravity work in the *Physics Letters A* paper. With that published theory, he did not see any real problems. On the other hand, as I sat in his office telling him about the ongoing second part of my work dealing with possible loops in time generated by two light beams circulating in opposite directions, he stopped me.

Guth said that he could see how loops in time could happen if I based this part of my theory on the model of the ring laser in my *Physics Letters A* article, wherein I had the light beams circulating in only one direction. But he didn't believe that two beams of light circulating in opposite directions would work. His intelligent critique struck a chord.

In the beginning of my recent work, I had been uneasy about the point he was now making. For one thing, I knew that with two light beams circulating in opposite directions, the frame dragging effect could potentially cancel out. For objects such as rotating black holes, the frame dragging effect and closed loops in time seemed closely aligned. Guth's comments brought me back to my original concerns. I would have to attack head-on the problem of calculating the strong gravitational field of a light beam circulating in only one direction. I thanked Guth and left Boston knowing that I had more work to do.

Struggling with an appropriate model to calculate the strong gravitational field of a light beam circulating in only a single direction, I considered how the light might be channeled. One method I looked at was an optical fiber, essentially a light pipe. Just as a water pipe is used to channel water, an optical fiber is used to channel light. Because of their capacity for huge amounts of information transfer, optical fibers have found wide-ranging applications, including telephone communications. These light pipes seemed tailor-made for my use. To increase the effect of a single circulating beam of light I conceived of wrapping an optical fiber around a hollow tube in a helical spiral to create a sort of cylinder of light. With this I felt I was beginning to approach a model I could deal with.

Interest in my research had been steadily increasing among students and faculty and it was suggested that I give a physics colloquium at my own university about my latest research. My talk to about 150 students and faculty members was similar to what I had given at Michigan with one important difference: I outlined the need for calculating strong gravitational field of a light beam circulating in a *single direction* for the creation of closed loops in time. The lecture was well received, and helpful comments came from my colleagues, as well as my former department head, Ralph Bartram. Feedback from respected colleagues was a powerful reason, I realized, for a scientist involved in ongoing research to make such presentations.

In my work on the weak gravitational field of a ring laser, I had considered a single neutron spinning like a little top at the center of the laser ring. Bartram, a solid-state theorist, suggested that I might consider a beam of neutrons. By sending the beam through the stronger gravitational field of a light cylinder, Bartram thought I might see an intensified frame dragging effect in which the entire beam of neutrons would be turned around as they moved along the cylinder. I felt that Bartram's suggestion might be worth trying at some point when we reached the experimental stage.

Unlike other scientific disciplines such as chemistry and biology, physics has a formal division of labor between theory and experiment. A theoretical physicist uses mathematical relations to explain a theory, while an experimental physicist sets up whatever apparatus is required and makes the actual measurements that constitute a physical test of the theory. To be successful, an experimental physicist must have a facility for designing sophisticated apparatus and taking precise measurements, while a theoretical physicist must be skilled in using mathematical techniques to explain the workings of nature. Traditionally, there exists a natural tension between the two disciplines. If a theorist believes he has the correct equation or theory, he is not thrilled by negative results of an experiment that disproves his work. In the world of an experimentalist, a negative result can be just as important as a positive one.

The prediction of a physical behavior of nature can first be made through an experiment and/or observation, but also, the theory for a physical behavior can come first. In the real world of physics, it works both ways about equally. An example of the former is the discovery of the photoelectric effect by experimentalist Philipp Lenard,[28] who observed that changing the color of light that is shined on a metal surface changes the speed of electrons that are ejected from the surface. Using quantum theory, Einstein later came up with the theory and equations for showing how the increasing frequency (color) of light gave more energy of motion to the electrons, work for which he won his Nobel Prize. An example of the latter is Einstein's mathematical derivation of the famous equation $E = mc^2$ that predicted that a small amount of mass (m) could be liberated to provide an enormous amount of energy (E). Einstein was never involved in experiments to prove his theory. In fact, although most theorists are interested in having their work verified, Einstein was largely indifferent to experiments to prove his theories. (For security reasons due to his German citizenship, Einstein was not allowed to be part of the most famous experimental verification of his celebrated $E = mc^2$ equation: the Manhattan Project.)

As a theoretical physicist, I learned long ago that I did not have the aptitude for experimental work. However, the time machine was too important for me to wait until experimenters decided they were interested in my theory. As a result, I was keenly interested in being closely involved in the experimental side of the project. I knew, however, that to adequately test my theory, I would need the help of an experimentalist who possessed specific scientific skills.

Such qualified assistance came from close at home in the person of an innovative experimental physicist, Chandra Roychoudhuri, research professor of physics at UConn who runs the university's privately funded research laboratory set deep in the woods several miles away from the main campus. Born and educated in India, Roychoudhuri graduated from Jadavpur University, Calcutta, then came to the United States. He received his Ph.D. in 1973 from the Institute of Optics at the University of Rochester in New York. Roychoudhuri had attended my UConn colloquium and afterward came up to me and commented about the "underlying symmetry" of my theory. He was familiar with the prediction from Einstein's general theory of relativity that matter has a gravitational effect on light, verified by Arthur Eddington in his famous 1919 observation of the bending of distant starlight by the gravitational field of the sun. Chandra, a specialist in the exciting new field of photonics, told me that he liked the notion that light also has a gravitational effect on matter. Photonics does for light what electronics does for electrons. Just as electronics is the control of the flow of electrons, photonics is the control of the flow of light. The science and applications of photonics are usually based on laser light.[29]

Chandra thought my theory broke new ground on light and matter interactions, and should be set up experimentally. He suggested a collaboration. Realizing that Chandra's background in photonics—and specifically lasers—made him an ideal collaborator to test my theories, I wholeheartedly agreed.

Chandra, of medium height and possessing a warm smile that he shows often, has a great sense of humor and also loves to get into long philosophical conversations. His lab is situated in what used to be a cafeteria and dining hall; the entire structure is painted institutional green. Lined up inside the expansive space are about a dozen optical tables, on which any number of different laser experiments are always being carried out for various private industries. Chandra is renowned for his work with high-powered laser diodes, which are tiny lasers (about .006 inches high which is only six times the width of a human hair.) that can produce peak power of 100 watts.

During regular meetings to talk shop, Chandra and I considered photonic crystals as a possible alternative to optical fibers as a means of channeling light along a helical path to form a light cylinder. The study of photonic crystals originated with the 1987 work of Eli Yablonovitch at Bell Labs in Homdel, New Jersey, and Sajeev John at the University of Toronto. A gemstone such as opal is an example of a naturally occurring photonic crystal. The particular iridescent color of opal is the result of the way light is channeled through the crystal structure of the gemstone. Armed with the knowledge that optical fibers or photonic crystals might be used to produce a light cylinder, I decided to use a circulating cylinder of light as the basis of my theoretical calculation.

Since it can be written in a single line, Einstein's gravitational field equations look deceptively simple. However, when taken out of the highly compressed tensor calculus notation, they represent a set of ten extremely complex equations. To do the calculations, I reached back to my experience of finding an exact solution for the Einstein field equations for the strong gravitational field of an evaporating black hole in an inflationary universe. In that case, it had been necessary for me to combine two solutions of Einstein's equations to form a new solution. I had combined the Vaidya black hole solution with the de Sitter cosmological solution to produce the Vaidya-de Sitter solution. My experience with this technique proved to be of significant value in dealing with the present problem.

I decided to dispense with trying to model mathematically either an optical fiber or a photonic crystal. Instead, for the sake of generality and to keep the light beam on a cylindrical path, I elected to use a geometric constraint. This constraint was represented by a static (nonmoving) line source. Light naturally wants to travel along a straight line. The only purpose of the line source in my calculations was to act as a general constraint to confine the circulating light beam to a cylinder. (Set up experimentally, the line source could look like wrapping a piece of string around a maypole, with the string being the light beam and the maypole serving as the line source.) The light beam itself would be conceived of as a massless fluid flowing in only one direction around the cylinder. This meant that the solution really contained two solutions: one for the circulating light and one for the static line source.

In March 2002, as I was still working on solving the field equations, I received an invitation to give a talk about my research at the June meeting of the International Association for Relativistic Dynamics (IARD) Conference in Washington, D.C., in three months. The conference was being hosted by Howard University and would be attended by experts in relativity theory from around the world. Excitedly, I accepted the invitation. This would be the first time my work would be exposed to specialists in my field. Since I had yet to finish my calculations, the opportunity presented was both exciting and nerve-racking. I now had a deadline, and the pressure was on.

In trying to solve Einstein's gravitational equations for the strong gravitational field of a circulating cylinder of light, the task was made even more difficult by the fact that the full Einstein gravitational field equations are highly nonlinear differential equations. Unlike linear differential equations, for which standard techniques of solving the equations exist, every nonlinear differential equation has to be considered on a case by case basis. In a linear equation, $2 + 2 = 4$. For a nonlinear equation, it can happen that $2 + 2 \neq 4$, wherein $\neq$ means "not equal to." Nothing could be taken for granted.

There have been many exact solutions of the Einstein field equations done by others. However, most of the solutions do not have a physical meaning; in other words, there is no identifiable source for the gravitational field of the equation. A few of the exact solutions with identifiable sources are the Schwarzschild solution for the gravitational field outside of a nonrotating spherical massive body, the Kerr solution for the gravitational field outside of a rotating spherical massive body, and the de Sitter solution for an expanding universe.

I was attempting to find an exact solution of the Einstein gravitational field equations for the gravitational field of a circulating light cylinder. Without the aid of a computer, I set about solving each of the ten nonlinear differential equations for the areas outside and inside the light cylinder. I spent every available moment I had attempting to solve these time-consuming equations, often working twelve to fifteen hours a day. Sometimes after working nearly all night I would drag myself to the university to give my lectures after only an hour or two of sleep. Simultaneously exhausted and charged, I had never worked so hard in my life.

After two months, I finally had an exact solution of the field equations. With great anticipation, I began studying the temporal properties of the solution, in other words, those that related to time. It didn't take long for me to notice there was a term in the solution that ordinarily looked like a circle in space outside the cylinder of light. When I worked the numbers and the frame dragging effect of a circulating light cylinder became strong enough, *the circle in space turned into a circle in time.* This meant that there were closed loops in time outside the circulating cylinder of light that would lead to time travel into the past.

"Closed loops in time . . ."

I put my pencil down and rubbed my temples.

The clock on my desk showed it was after 3:00 AM.

I now had a completed theory for time travel to the past. Turning out the lights, I went to bed.

*   *   *

As the airliner descended toward Ronald Reagan International Airport in Washington, D.C., on June 25, 2002, I peered out the passenger window at our nation's capital. Notwithstanding a degree of anxiety, I felt ready to present the results of my work to the annual meeting of the International Association for Relativistic Dynamics.

Once inside the airport terminal, I walked over the covered bridgeway to the Metro station. After transferring to the green line, I got off at the Shaw-Howard University exit, turned north, and walked six blocks to the main campus of Howard University. It seemed strangely coincidental yet wonderfully appropriate that this august international group which convened at different worldwide locales each year (the previous year at Tel Aviv University in Israel) was having its meeting this year at Howard University, with its rich tradition of providing educational opportunities and excellence to African-Americans since shortly after the Civil War. This year—my year, *and here of all places*—I would stand before some of the world's most respected physicists to present the whole of my work and life's ambition. I would be one of three African-Americans in the room; another, Tepper Gill, was a Howard University faculty member.

Inside the lecture hall I looked up at the rows of occupied seats that rose upward from the podium. Some of the biggest names (as well as brains) in my field awaited me. After catching my breath following the joke about having no more than sixty transparencies to show (I had twenty-six), I began by explaining how my theories were solidly based on Einstein's general relativity theory.

I went over the details of how closed time loops were formed in the gravitational field circulating cylinder of light. Pointing to relevant equations projected on an overhead screen, I showed that the same term that led to frame dragging in space was needed to generate the closed loops in time. "The circulating light beam causes a swirling of space that produces a swirling of time, which in turn results in the formation of closed time loops."

In the course of my presentation, I projected on the screen various illustrations, equations, and final solutions, all for the purpose of showing that space and time can be manipulated in a whole new way that would led to the possibility of time travel into the past.

The final transparency, however, was something different. It was of a photo taken long ago of my family at the Bronx Park with my smiling handsome father holding my little brother, Jason, while I stood next to my beautiful mother. (My younger brother, Keith, had not yet been born.)

I concluded the talk by telling the group about my original motivation for being interested in the problem of time travel, explaining how after my father's death I was inspired by H. G. Wells' book to one day build a real time machine.

"My original motivation for everything—learning math and science, going to college, becoming a physicist—was so I could see my father again."

Then the lecture was over.

There was a silence from the audience that lasted somewhat longer than I would have liked. I wasn't sure how to take it. In retrospect, I believe the group was stunned by the sudden personal turn my talk had taken, something that was not often heard at such scientific gatherings. Then, the applause came.

As I expected, some perceptive questions followed. One had to do with the circumstances under which the frame dragging effect could be observed. I explained that my experimental colleague, Chandra Roychoudhuri, was presently studying practical designs for the implementation of neutron-beam experiments to test the gravitational frame dragging effect.

When there were no more questions, Bryce DeWitt, the legendary cofounder of the quantum theory for gravity whose presence at my talk had left me both anxious and awestruck, came to his feet.

I still had amongst my books one by DeWitt, *Relativity, Groups and Topology,* which was well-worn and had helped me years ago with fundamental understandings crucial to the completion of my Ph.D. thesis.

I didn't wait with bated breath so much as I refused to breathe.

"I don't know if you'll ever see your father again," DeWitt said, his eyebrows raised as he peered intently at me. "But I do know he would have been proud of you."

DeWitt's comment had a profound impact on me, and not only because it was said by a man whom I greatly respected. I had been concentrating so much on close-up equations that I hadn't taken time to see the wider focus. My lifelong thirst for knowledge, my science education, my university teaching profession: yes, Dad would have understood my drive to better myself, and he would have been proud of me. In a real sense, he had been the one–in the short time we were together in life–who put me on this path. Even without solutions on my pad of paper, I knew that as surely as I knew anything in the universe.

I began to feel whole for the first time in many years.

* * *

The second half of my time travel theory was published in the journal *Foundations of Physics* in September 2003. Entitled "The Gravitational Field of a Circulating Light Beam," the paper contained the new exact solutions of Einstein's gravitational field equations for the gravitational field of a circulating light cylinder and showed that closed loops in time resulted.

Another part of DeWitt's comment after my talk turned out to be true.

As my results indicated, when I turned off the light flow, the time loops disappeared while the line source around which the light had circulated remained. The closed loops in time had been produced by the circulating flow of light, and not by the non-moving line source. When it came to translating my theory into the design for a circulating-light time machine, it was clear the circulating flow of light would be the machine's on-and-off switch. Per my results, in order to achieve closed loops in time, the light source had to be on.

Closed loops in time would continuously be formed only as long as the flow of light continued. These loops would stack up on top of each other and be connected to form a spiraling helix that would look rather like the familiar child's toy, the Slinky. Even as the earth moved in space, the time loops of the helix would adjust to the new location. If the circulating flow of light was left on for a year, then by entering the helix a year from now, someone would be able to spiral back along the time loops of the helix to as far back as the first day a year ago when the circulating flow of light was turned on.

As the various implications involved in the design and operation of the time machine began to sink in, a realization hit me: *My time machine could only carry a time traveler back to the moment the machine was turned on, and not one second before.* Why I didn't see this sooner, I do not know. Perhaps it was again a matter of being too close to the equations to see the big scheme of things. What it meant was that when the first time machine that could transport a human became operational, our descendants might be able to visit us, but we could never visit our ancestors.

It would not be like the time-travel movie *Frequency*, which I had seen when it was released in 2000, and then bought when it came out in video, and saw again—and again. Next to *The Time Machine*, it had the most impact of any film in my life, touching on the essence of what my dreams were about. *Frequency* tells the story of a police officer named John Sullivan who is able to communicate across time with his dead firefighter father, Frank Sullivan (played by Dennis Quaid). John's father had died thirty years earlier fighting a fire in a warehouse. During an evening of unusually intense solar flare activity, John turns on his deceased father's old ham radio set. After an exchange with an unknown ham operator, John becomes aware that the man is his father. John realizes that he is in a position to alter his father's fate. He warns Frank about the fire he has yet to fight, and saves him. In so doing, John changes his own life by having his father present for his childhood and beyond.

I would never get the opportunity to do the same.

I would not be able to use my time machine to see my father.

The grown man and scientist I had become could now let go of the final emotional vestiges of his shattered childhood. My father was gone, and there was nothing I could do about it other than live my own life with pride and courage, and fill as many days as I had left with valued people, times, and works.

I thought back to what J. Robert Oppenheimer had said when he was asked years after World War II about the motivation that led to the successful completion of the Manhattan Project. Surprisingly, it wasn't to build an atomic bomb to end the war. Oppenheimer said the scientific project had been "technically sweet," and claimed that had been the true motivation for many of the scientists who did end up building a new kind of bomb that ended the war.

Even though my original goal of being able to travel back to the 1950s had morphed into something else entirely, I was highly motivated to complete the project out of sheer curiosity and the sweet science of seeing the device work. I now began to plan how I might turn my theory for breaching the space-time continuum into a practical reality, with my overriding hope and intention that it would be used only for peaceful purposes, never violent, war-like endeavors.

It was time to build an experimental time machine.

Thirteen

Building the Machine

⧗

Theoretical physicists are generally not concerned with practical applications of their work. Applications that lead to patents are normally reserved for specific technological devices; theories cannot be patented.

I had, however, a somewhat different take on this after reading *How the Laser Happened,* the autobiography of Nobel laureate physicist Charles H. Townes.[30] In his book, Townes indicated that it was not necessary to have a working model of a device in order to have it patented. He suggested that a researcher with a concrete theory as to how a new device will work should go ahead and apply for a patent.

I consulted with UConn's Center for Science and Technology Commercialization. The director of technology licensing suggested that I file a provisional patent that would establish a record of my idea while giving me time to work out a full patent, something I could wait to do until the circulating light-gravitational frame dragging experiment began in earnest. (A provisional patent application does not require as much information as a full patent, nor does it cost as much money to file. Also, it lasts for only a year.)

My goal was not to be secretive or possessive when it came to my time travel research. In fact, quite the opposite was the case. Historically, once an original theory is published in the scientific press, it is common practice for other scientists and engineers to use it in their own work. That is how science advances. While Einstein is credited with the theoretical discovery of the stimulated emission of radiation that is behind the laser, other researchers developed the first working laser. I am sure Einstein would not have had a problem with that had he been around to witness the dawning of the laser age.

In studying U.S. patent guidelines, I realized it would not be possible to patent a time machine as such. However, I could request a patent for an application in which the time machine was a central component. After giving this further thought, I came up with the idea for a device called LOTART. In my patent application filed July 2, 2003, with the U.S. Patent and Trademark Office, I provided the following information under "Detailed Description":

A laser optical time machine and receiver transmitter (LOTART) is a communication device consisting of a unidirectional circulating light beam connected with a signal receiving and transmitting device. The time machine receiver would be capable of receiving long-range signals from an external transmitting device constructed for a particular application at a specified time and place in the future. The time machine internal transmitter would then send signals to an earlier moment along closed time lines with information about subsequent external conditions. For example, if at some later time a planetary space mission is successful, then a signal could be transmitted from the landing module to the earthbound light cylinder time machine. The received signal would then be transmitted within the light cylinder time machine to the present. The reception of the signal would indicate, at the present time, whether the future mission was a success. Based on the receipt of the signal, it could be determined whether or not it was necessary to change mission parameters. This could lead to considerable savings in terms of cost and manpower of space exploration.

The proposed optical time machine consists of a laser beam sent through a unidirectional cylindrical waveguide in an appropriate medium connected to a receiver and transmitter. The resultant circulating light beam generates a gravitational field that contains closed time lines. A signal receiver-transmitter is located in the region outside the circulating light cylinder. Specific electromagnetic signals detected at the receiver would then be transmitted along the closed time lines from a designated later time to the moment when the LOTART was turned on.

In the "Claims" section, I indicated that LOTART was:

* A method of generating the closed time loops associated with gravitational field of a circulating light beam and the reception of signals sent at some future time to be transmitted to the present.
* A method of forming a unidirectional cylindrical light configuration in an appropriate optical medium. The cylindrical configuration can be approximated by photonics crystals, optical fibers, or a stacked array of unidirectional ring lasers.

The following diagram, captioned "Light cylinder (LC) wave

guide and signal receiver transmitter (SRT)," was submitted with the application:

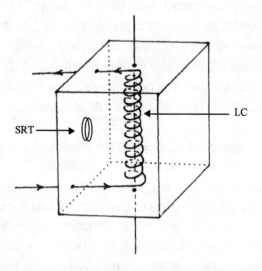

Fig. 1 Light cylinder (LC) wave guide and signal receiver transmitter (SRT)

The government granted a provisional patent for LOTART in August 2003. Although it had been an interesting exercise to speculate about possible applications in order to qualify for a patent, the most pressing mission was to determine experimentally whether a particle of matter or signals of some type could be sent back in time as predicted by my calculations.

From the outset of our collaboration, my experimental colleague, Chandra Roychoudhuri, and I decided to initially concentrate on demonstrating the first part of my time-travel theory: that the gravitational field of a circulating beam of light would lead to a twisting of space with the resultant frame dragging. There were practical as well as theoretical reasons for this decision. On the practical side, all that was needed for frame dragging was a weak gravitational field. In contrast, a strong gravitational field was required to produce closed time-like loops. Since less energy was required to produce a weak gravitational field, it made sense to first look for frame dragging—an easier experiment to set up and a necessary condition for time travel to occur.

From a theoretical standpoint, the equations showed that closed loops in time were generated at higher energies by frame dragging. If we could not experimentally produce frame dragging, it would be pointless to look for closed loops in time.

The machine would be built in phases. The first phase would be to design equipment to verify that a circulating beam of light at low energies *did* twist space as I predicted. Only when space twisting was observed by means of frame dragging would the second phase commence. The second phase entailed going to higher energies to create closed loops in time.

To begin the first phase we had to develop a practical design for the ring laser, and use it on an appropriate test particle. The particle we chose was the neutron, a subatomic particle that makes up (along with protons) the nucleus of most atoms. A neutron has a property called "spin" that would make it useful in the experiment. Any rotating object is said to be spinning. The earth, for example, is rotating around an axis through the North Pole. Therefore, the earth can be said to be spinning about its axis. The neutron also has an axis about which it rotates or spins. Normally, the direction of the neutron spin doesn't change. However, my theory showed that if the neutron is placed at the center of a ring laser, the gravitational field will cause the spin direction to be dragged around. The change in the direction of the neutron spin is called precession. A key result I had derived from my calculations was an equation for the frame dragging precession of a spinning neutral particle such as a neutron:

$$\Omega = \frac{8\sqrt{2}G\rho}{ac^3}$$

In this equation, Ω represents the rate at which the neutron spin direction changes. The equation contains two fundamental constants of nature, G and c. G is called the universal gravitational constant and c is the speed of light. There are two variables, P and a, in the equation that can be changed according to the experimental setup. P is the intensity of the laser beam and the variable a is the length of any side of the square that makes up the ring laser. The equation shows that increasing the intensity of the laser or decreasing the ring laser size will increase the rate at which the spin direction of the neutron changes. This means that the experiment should be designed to achieve the largest possible intensities for the smallest possible ring size.

In my original ring laser-gravity paper in *Physics Letters A* I had used mirrors to establish a circulating light flow. This configuration had severe limitations since, on a practical level, the size of the laser ring is limited by the size of the mirrors. Mirrors could be made only so small. Chandra and I realized that something else was needed. We came up with an alternative scheme to use four separate but intersecting laser beams to form a square. The symmetrically focused laser beams from the four corners of the square would simulate the effect of a high-intensity ring laser. The intersecting beams could be adjusted to achieve extremely small ring sizes of about .000001 meters (.00004 inches) in length.

It also became clear that the effect of the ring laser could be further increased by stacking the rings atop each other. The final configuration would be a laser light tower with each level consisting of four intersecting laser beams. The laser beams of each level of the tower would be provided by high-powered diode lasers. Each diode laser can produce 10 watts and is about .000155 meters (.006 inches) high. The circulating light tower would produce a space twisting gravitational field. My calculations showed that an observable frame dragging effect would be seen for a tower stacked with 10,000 diode lasers, which would lead to a total tower height of about five feet.

To see the space twisting, frame dragging effect, a beam of neutrons would be shot through the center of the tower. As the neutrons emerged from the tower, the change in their spin direction would be measured. If the spin direction changed by the amount predicted in the equation, this would prove that the neutrons had experienced frame dragging due to the space twisting effects of the gravitational field of the circulating light tower.

It is important in scientific experiments to consider different possible ways to achieve a desired effect. It turned out that there was another way of observing the gravitational frame dragging effect of the ring laser that did not involve particles of matter at all. This approach occurred to me when I considered some of my earliest research, in particular what I had done with my first graduate student, Fred Su, who had studied the frame dragging effect caused by the gravitational field of a rotating black hole. This was done by looking at what happens to a light ray as it passes near a rotating black hole.

A fundamental property of light is that it vibrates up and down as it moves forward. The plane formed by the up and down vibration is called the plane of polarization of the light ray. In Fred's thesis, he showed that the plane of polarization of the light ray is twisted by the gravitational field of a rotating black hole. I realized that a similar twisting of the plane of polarization of a light ray should happen if we sent the ray through the gravitational field of the circulating light tower. We now had two different possible experimental arrangements that could test the first phase.

As Chandra and I were considering these various experimental strategies, there was news about an exciting satellite experiment underway. *Gravity Probe B* was set up to test the gravitational frame dragging caused by the rotation of the earth. The $700 million project was a collaboration between NASA and Stanford

University, with NASA providing the money and Stanford serving as the prime contractor for the mission and responsible for the design and integration of the science instruments, as well as for mission operations and data analysis. The principal investigator for the experiment is Stanford physicist C. Francis Everitt.[31] I thought it might be valuable for me to meet with Everitt to learn more about his frame dragging experiment, and tell him about mine. Everitt agreed to a visit, and I flew to California in June 2004.

This was my first visit to Stanford, located in northern California about forty miles south of San Francisco, and I was struck with the beauty of the tree-lined campus. Shown to Everitt's cluttered office, my first impression of him was that he resembled photographs I had seen of a sixtyish Albert Einstein. Everitt is soft-spoken, with a British accent, and a mop of stringy hair that falls to his shoulders. After some pleasantries, he gave me a tour of the laboratory complex.

Everitt took me first to a model of the cryogenic probe that housed the four gyroscopes so essential to the experiment. The probe looked like a giant stainless-steel thermos bottle. Each gyroscope was a perfectly smooth 1.5-inch sphere—about the size of a Ping-Pong ball. "These spheres are the roundest objects ever made by man," Everitt explained. The tiny spheres were enclosed inside a housing chamber to prevent disruption from sound waves, and chilled to almost absolute zero to prevent their molecular structure from creating a disturbance. Everitt claimed the accuracy of the gyroscopes to be "thirty million times greater than any gyroscope every built."

The experiment was not spur of the moment, but had a long history.

In 1918, three years after Einstein had published his general theory of relativity, two Austrian physicists, Joseph Lense and Hans Thirring, predicted that a massive rotating body—such as the earth—would drag the empty space surrounding its mass. This effect was dubbed frame dragging. By 1960, however, frame dragging due to a massive rotating body had still not been observed.

It was at this juncture that Stanford physicist Leonard Schiff had an idea for a possible way of observing frame dragging due to the earth's rotation. He suggested that a gyroscope could be placed in polar orbit to measure the effect.[32] A gyroscope is simply a rotating or spinning object. Sans any forces acting on the gyroscope, the spin direction doesn't change. If Einstein's general theory of relativity is correct, once the gyroscope is in a polar orbit the rotation of the earth will result in a twisting of space around the satellite. This twisting of space would change the spin direction of the gyroscope as it is dragged around in the direction of the rotating earth. There was one problem: at the time Schiff proposed this experiment, the technology needed to carry it out did not exist. Not only was it a matter of not having the precise scientific and measuring apparatus required, but the U.S. had been in the space business only a couple of years. Rockets and orbital satellites were still an emerging technology.

Later in the twentieth century satellite launches became routine, and advances in superconductivity (electrical currents that occur with no resistance) and materials technologies—especially the ability to produce perfectly spherical spinning gyroscopes—allowed for the possibility of an experimental test of gravitational frame dragging by the rotating earth. Unfortunately, Schiff died in 1971 before the experiment could be set up. In 1981, Everitt became the principal investigator for the satellite experiment.

Two months before my visit, on April 20, 2004, a 21-by-9-foot satellite carrying the probe containing four gyroscopes was successfully launched by a Delta II rocket from Vandenberg Air Force Base. Everitt explained that final calibration of the gyroscopes were, as we spoke, still being done by sending signals from the Stanford control center to the satellite in orbit. After that was completed, the experiment would begin. If Einstein's predictions were correct, the four ultra-precise gyroscopes should detect that small amounts of time and space are missing from each orbit. To measure each orbit, the gyroscopes are aligned with a guide star using a tracking telescope. A magnetic-field measuring device records the changes in respect to the guide star.

When we returned to Everitt's office, I told him about my own work, and my prediction of frame dragging due to the gravitational field of a circulating light beam. I had sent him copies of my two published papers, and I now asked his opinion of the experimental arrangements that Chandra and I had been considering to test my theories. Everitt thought that the most likely success would come with considering the frame dragging effect on the plane of polarization of a light ray that was sent through the gravitational field of the ring laser. The reason for this was that light was far easier to control than neutrons. The light ray could be reflected multiple times through the ring laser, which would considerably amplify the frame dragging effect on the plane of polarization of the ray and make the effect more measurable.

Before leaving I thanked the gracious Everitt for his time and learned opinions, and invited him to give a physics colloquium at UConn the following year. Everitt said that he had never been in New England during the fall and he had heard that the foliage was colorful that time of year. We set a date for his visit in October 2005. According to Everitt, sufficient data would have been collected by then to check on the validity of the prediction of frame dragging due to the gravitational field of the rotating earth. Until then, I would be eagerly awaiting, like everyone else in the physics community, those results.[33]

When I returned home, I discussed what I had learned with Chandra. He felt that Everitt was correct in his belief that greater measurability of the gravitational frame dragging effect of the ring laser would be achieved by observing the effect of the plane of polarization of a light ray. Nevertheless, we agreed that the best strategy for the present time would be to continue looking at both neutrons and light rays.

In our earliest discussions, Chandra and I both acknowledged the paramount need of funding for our project. Research funding can come from government, business, nonprofit and private sources. Government funding can be acquired by sending a proposal to an appropriate agency. Sources of government funding can be classified as nonmilitary and military. Examples of nonmilitary funding sources are the National Science Foundation (NSF) and the National Aeronautics and Space Administration (NASA). Military funding is connected with such organizations as the Defense Advanced Research Projects Agency (DARPA). I had, in fact, been contacted by someone with connections to DARPA. However, I felt the need for caution over the issue of receiving possible funding support from the military. My unease came partly from the somewhat complicated history of the laser.

In 1953, Charles Townes invented a device for amplifying microwave radiation that he called the maser (*m*icrowave *a*mplification by the *s*timulated *e*mission of *r*adiation). By 1957, Townes was thinking about how to build a device to amplify light based on the maser principle. Townes called the proposed device an optical maser. At the time, Townes was a professor of physics at Columbia University. Working independently, a Columbia graduate student named Gordon Gould began thinking about a device that he called a laser (*l*ight *a*mplification *s*timulated by the *e*mission of *r*adiation). In 1958, Gould left his doctoral studies to work for a small company named Technical Research Group (TRG). He was able to interest the company in his idea for a laser. TRG was able to obtain a sizable military contract to work on the laser from the Advanced Research Projects Agency (ARPA), the precursor to DARPA. Unfortunately for Gould, security restrictions were attached to the military largess, and Gould was unable to obtain the necessary security clearance.

His research notes were classified, meaning he could not even write anything for publication. Gould was shut out of his own project. While Townes' contribution to the development of the laser is well deserved, equal credit should have been given to Gould and no doubt would have been—perhaps along with a share of Townes' Nobel Prize—had it not been for the intervention of the military.

There are other examples in which the military—for reasons professed to involve national security—had taken a project away from a scientist. In a case I knew about, a laser researcher lost his project when the military saw the potential for a laser weapon to use against incoming missiles and aircraft and took it over. I did not want that to happen to this project. The last thing I wanted was to accept military funding only to have the work classified and snatched away—perhaps at the moment that the military realized the experiment worked. Chandra, who had also heard outrageous stories of military interference in science, agreed. Although military funding for scientific projects is available in large amounts, we decided early on not to seek or accept it.

All major scientific and technological endeavors require sufficient funding to carry out the actual experiments, and certainly Chandra and I were in the same position when it came to building the circulating light-time machine. A budget had been developed at UConn for the first phase of the time machine experiment, which had officially been designated as the "Space-time Twisting by Light" (STL) project. The budget for the initial phase—to demonstrate the gravitational field of a circulating light beam causes a twisting of space—totaled $286,000, which included the cost of personnel (postdoctoral and graduate students), equipment, research travel, and supplies. The University of Connecticut Foundation, Inc., a nonprofit institution, has established an account to receive contributions and manage funds for the project.[34]

In the meantime, I began considering possible strategies for carrying out the second phase of the time machine experiment: to establish the presence of closed loops in time due to the gravitational field of circulating light. Quite unexpectedly, a suggestion for an experimental test came as a result of another collaboration I was involved in. Often, physicists find themselves working on more than one research project, and that has usually been the case for me.

For a number of years I have had an interest in one of the major problems confronting astrophysicists: the so-called missing matter problem. That something was not quite right in the universe was first noticed by observers of the motions of stars in galaxies. One way that astronomers have of finding out how much matter is in the galaxy is simply by counting the number of stars in a galaxy. It is then simply a matter of arithmetic to add up the number of stars and determine the total mass of the galaxy. Another way of determining the mass of a galaxy is by looking at the motion of stars that are orbiting near the edge of a galaxy. It turns out that by observing the velocity of an orbiting star it is possible to calculate the mass of the galaxy about which the star is orbiting.

A shock came to astronomers when they determined the mass of a typical galaxy by first counting the stars making up the galaxy and comparing that mass with the mass of the galaxy as determined by observing the motion of stars orbiting around it. To their surprise, the two results did not agree. The number of stars counted gave a value for the mass of the galaxy that was much less than the mass that was determined by looking at the motion of stars. In a nutshell, this is the missing matter, or missing mass, problem.

A number of theoretical proposals had been suggested to resolve this problem. Since we count stars by seeing them because they are luminous, the most obvious answer is that the majority of the matter that makes up a galaxy must be nonluminous, meaning we can't see all the matter. Because most of the matter is nonluminous, it is called "dark matter." Some of this nonluminous matter could be in the form of black holes, also invisible.

A friend and colleague of mine, Mark Silverman, a British-born professor of physics at Trinity College in Hartford, Connecticut, became interested in the missing matter problem. Mark and I first met in 1987 at a conference that was held at Kings College in London to celebrate the hundredth anniversary of the birth of Erwin Schrodinger, who discovered the wave equation for the electron in 1926. Mark received his Ph.D. from Harvard in 1973 and is internationally known for his work on atomic and optical physics. He is a friendly, dapper man of medium height who sports a rather roguish moustache and short beard.

Mark thought that the missing matter might be due to a universal fluid made up of lightweight particles of a special nature. We decided to collaborate because it appeared that the properties of these particles could be related to earlier work I had done on Einstein's famous (or infamous) cosmological constant, in which he attempted to show—erroneously, as it turned out—that the universe was not expanding. The approach Mark and I have taken to the missing matter problem is promising, and we have published a number of papers discussing our results.

During one of our meetings, I told Mark about my work involving the gravitational field of a circulating light beam, and the formation of closed loops in time. Mark made a brilliant suggestion about one possible way of testing for the presence of the closed time-like loops. His original remarks involved using radioactive elements, which have proscribed life spans. Later, I modified his suggestion and applied it to the decay of unstable elementary particles.

The test would be set up as follows: consider that a circulating beam of light has generated a closed loop in time. Also: unstable elementary particles of a particular type all have exactly the same lifetime. Suppose that a beam of such particles is sent past a region that contains a closed loop in time. Those particles that approach the loop on the right side will be experiencing a different direction of time than those particles that approach the loop from the left. If a detector is placed on the other side of the loop, the lifetime of the particles that arrive at the detector will be different—depending on their path in time—as they encountered the closed loops in time. This should give a precisely measurable determination of the presence of closed time-like loops.

This is the kind of definitive result that would convince other physicists that a circulating beam of light had indeed created a closed loop in time. If we can send an elementary particle back in time, we can send any other matter back as well, although to do so would require more lasers and more energy.

When we are in the final stages of building the time machine, it is this kind of experiment that Chandra and I will first attempt to perform at his lab. If a difference in the lifetime of the decay of unstable elementary particles is indeed observed, then the concept of a circulating light time machine will have been verified—and the age of time travel to the past will be ushered in.

Fourteen

Time Travel Paradoxes

⧗

That my calculations predict time travel is possible only back to the moment the first working time machine is activated solves a problem raised by Stephen Hawking and others. Their question: if travel into the past became possible at some future date, why haven't we received time travelers from the future?

This then is one possible answer: We may not have seen time travelers because the first working time machine has yet to be switched on. Once a machine that can transport humans is activated, we may indeed start receiving visitors from the future, as they will then have a portal through which they can reach us. Well before then, however, we may begin receiving messages from the future via a time machine that can send and receive more elemental (nonhuman) forms of matter, such as radio signals.

There is a potential way around the limitations on time travel that my work predicts, but it would require the existence of intelligent life on other planets. There is every reason to believe that the universe is teeming with life. If it is not, and if earth turns out to contain the only forms of intelligent life, then a whole lot of space has been wasted. The SETI (Search for Extraterrestrial Intelligence) project has for decades been searching the sky for radio signals that might indicate that intelligent life, other than our own, exists in the universe. To date, no interstellar radio signal has been found that can be unambiguously interpreted as a message from an extraterrestrial civilization. Nevertheless, the search continues.

An even more promising direction for finding life on other planets has come from the discovery of extrasolar planets. Beginning in the 1990s, great excitement was generated when astronomers confirmed the observation of planets outside our solar system that were orbiting stars very much like our sun. The discovery of these extrasolar planets was made possible by greatly improved telescope technology. On October 6, 1995, two Swiss astronomers at the University of Geneva, Michel Mayor (born 1942) and his graduate student, Didier Queloz (born 1966), announced the first definitive observation of an extrasolar planet, a planet which orbits a star other than our sun and therefore belongs to another planetary system. The planet—named 51 Pegasi b—is located in the constellation of Pegasus about 47.9 light-years from earth. (One light-year is about six trillion miles.)

Extrasolar planets are too small to be seen directly; indirect methods of observation are necessary. One method of detection of extrasolar planets is by observing the gravitational influence that the orbiting planet has on the star. Another way of observing an extrasolar planet is the transit method: as a planet crosses in front of a star, the shadow of the planet causes the light of the star to dim. By early 2006, nearly 200 extrasolar planets had been detected. Most of the extrasolar planets are much larger than the earth (around the size of Jupiter, which is 316 times more massive than earth). However, on January 25, 2006, a planet no more than five times the mass of the earth was found. Astronomers believe it is only a matter of time before an earth-like planet is detected. The consequences of such a discovery would be profound. Our galaxy alone contains about 100 billion stars. If even a small fraction of these stars contained earth-like planets, it is possible that extraterrestrial civilizations which are both less and more advanced than our own do exist.

I do not discount the possibility that an alien world exists with technology so advanced that time travel has become a part of their civilization. It is possible that one distant day, when we have developed space propulsion technology advanced enough, we will reach one of these worlds.[35] Suppose they built and turned on a time machine a few thousand years ago. Imagine the excitement of seeing ancient Egypt, Greece, Rome and witnessing historic events we have only read about in history books.

Regardless of the method of travel, going backward in time opens up a Pandora's box of troubling paradoxes. One of the objections that has been raised against time travel into the past, by both philosophers and scientists, is that it could lead to serious contradictions. The chief example is the so-called Grandfather Paradox. This paradox can be illustrated by a parable. Consider if you will the story of one Keith Fraser.

On October 31, 2050, Ted Fraser, a bright and ambitious young man, is fortunate enough to have an interview with a top legal firm in his hometown. The senior partner of the firm is immediately taken with Ted's qualifications and personality. Ted is hired and eventually meets the senior partner's attractive daughter, Emily. Ted and Emily fall in love and are married. In 2052, the happy couple have a son, David. Ted passes away in 2074. A few years later, David marries and has a son, Keith. As Keith grows up, David constantly tells his son about the wonderful grandfather that he'll never know. In 2095, Keith joins a top-secret government project that had developed advanced time travel fifty years earlier. Obsessed throughout the years with wanting to have known his grandfather, Keith volunteers to go back in time. He is transported to his grandfather's hometown on the morning of

October 31, 2050. Spotting his grandfather, whom he recognizes from old photographs, Keith hollers to get his attention. Ted, taken by surprise, turns suddenly, trips, and breaks his leg. Taken to the hospital, Ted misses the job interview with the law firm, and the job goes to someone else. At this point in the story we arrive at a major paradox. By missing the interview, Ted never has the opportunity to meet the boss's daughter. As a result, Ted and Emily never have a son, David. Since David is never born, our time traveler Keith is never born. But if Keith is never born, then how can he travel back in time to change his grandfather's life?

This is the crux of the paradox: how is it possible to travel into the past and do something that could wipe out your own existence? There have been many attempts to deal with this paradox. One way to eliminate such time travel paradoxes has been to say that nature conspires to prevent such an unnatural changing of history. This "prevention by nature" is related to Hawking's Chronology Protection Hypothesis, which conjectures that the laws of physics will somehow prevent the successful operation of a time machine. For example, when a time machine is turned on, it will somehow destroy itself before any history-altering events are allowed to take place.

Another resolution of the Grandfather Paradox is nearly as strange as the problem itself. This resolution is based on the other main pillar of twentieth century physics: quantum mechanics. Unlike the classical world of Newton, which deals with certainties, the world of quantum mechanics deals with probabilities. One application of quantum probabilities is the parallel-worlds theory of the universe, which states that at every possible decision point, the universe splits into different parallel branches (i.e., the cheeseburger or tuna sandwich dilemma). It's important to remember that this splitting does not refer to conscious choices—it just happens, and regardless of which universe we live in we would not know about the existence of the other.

The application of parallel-worlds theory to time travel was developed by Oxford physicist David Deutsch, and we can see how this theory leads to a resolution of the Grandfather Paradox by reconsidering the Keith Fraser story. At the moment that Ted's grandson Keith arrives in the past, there is a split in the universe. This parallel universe is different from the universe that Keith originally left. It is in the new parallel universe that Keith arrives to meet his grandfather. Even though Keith precipitates the events that led to his grandfather's missed opportunities, it does not lead to a paradox because Keith is in a new parallel universe and nothing he does in the new universe affects the old universe. In the old universe, his grandfather makes the interview, meets his future wife, and eventually has a grandson named Keith. As bizarre as this parallel-universe scenario sounds, it is not ruled out by the laws of physics.

In addition to chronology protection and parallel-worlds, there is another intriguing possibility for the resolution of time travel paradoxes. Suppose that a suitable encrypted transceiver receives a message from the future. The only physical reality that we have is the present moment. If we act on the message, then we will have chosen a particular future. This new future is likely to be different from the future from which the message was received. In other words, we really will have changed the future. That said, there is a fundamental uncertainty associated with our decision to act on any information we receive from the future. How can we be certain whether the information received is from our future or the future of a parallel universe? Consequently, even with a working time machine, there exists the potential for a degree of uncertainty.

Science-fiction movies have sometimes dealt with the perils of changing the past without having knowledge of the consequences for the future. One of the more recent films that considered the consequences of altering the past is the 2004 science-fiction drama, *The Butterfly Effect*. The story is about a young man plagued by unhappy childhood memories. He decides to change the past by using his ability to travel back in time. The problem is that every time he alters the past, the future consequences become increasingly worse.

Unlike traveling into the past, time travel into the future does not lead to such paradoxes. This is because once you arrive in the future, there's nothing you can do to change the past. You can only live with the consequences of what has happened during the time you were traveling to the future.

As for traveling into the past, I believe it will ultimately happen. When time travel does occur, abuses will naturally arise. As with any new, powerful technology, time travel will have to be regulated to prevent ill-usage. It will be up to society to ensure that time travel is used for the benefit of the human race.

The question of whether time travel occurs between parallel worlds, only in this world, or whether nature somehow will intervene to prevent time travel can only be answered after the first time machine is turned on. I believe that someday mankind will be able to answer the question "What happens when we go back in time and change the past?"

Time travel could allow us an unprecedented control of our destiny. Ultimately, however, the only thing any of us really have is the present moment.

*   *   *

If Einstein could come back for an hour and we could sit on a park bench and talk, I wonder what I might say to him. I think I would start by telling him that he wouldn't believe what we now know about the universe based largely on his work, for instance, the Big Bang theory, and how we know that the universe started out as a cataclysmic event. I'm sure he would also want to know the answers to some questions he wasn't able to solve.

"You were right, a unification of the forces of nature is necessary, but the reason you didn't solve it with just two forces—gravity and electromagnetism—is that we now know there are four forces at work. The strong nuclear force and the weak nuclear force make up half of the unification solution that governs all the interactions in the universe."

I would also tell him how a satellite is circling earth now to prove his theory that a large mass causes frame dragging. I would explain my own theory that shows the gravitational field of a circulating light beam also causes frame dragging. "We are understanding more deeply what your general theory has to say about nature of space and time." I would tell him that the gravitational field of circulating light is not only going to cause a frame dragging effect, but that according to my calculations it could also lead to closed loops in time, which means the potential for time travel into the past.

Do I think Einstein would sign on for my time travel theory?

Well, I don't believe he would take it for granted. I think he would say, "Show me the calculations."

I would be interested in knowing if Einstein read the same book that had so motivated me: *The Time Machine*. I would like to think that he did. Indeed, at a 1930 banquet in London where Einstein and H. G. Wells were both present, Einstein at the dais acknowledged Wells, saying he was pleased to see the novelist "whose views of life I'm particularly attracted to."

Before our hour on the park bench was over and Einstein returned to being one for the ages, I'd surely say, "Dr. Einstein, thank you for your general and special theories of relativity, and for all your other unbelievably incisive work for all of mankind. Thank you, too, for being my lifelong inspiration."

*   *   *

My mother is still alive and going strong at age eighty-two. When I visited her a few months ago, she took me to her church, Mount

Zion Baptist, in Altoona. She was beside herself with pride in showing me off to the other church members, introducing me as "My son, Doctor Ronald Mallett."

In truth, the loss of my father at an early age could have badly derailed me. As it was, I became withdrawn, and a habitual truant. Were it not for my dream of one day building a time machine, I might have dropped out of school and taken a wrong path in life. "My dream," I recently told an audience of young people, "helped keep me out of the state pen, and got me into Penn State."

My mother and father's other sons were driven to succeed, too. My younger brother, Jason, always a good people person, was a member of the renowned IBM sales force, and retired as an executive. My youngest brother, Keith, always a doodler, is a well-known commercial artist and portrait painter. My sisters, Eve and Anita, work as a computer graphics artist and medical assistant, respectively.

On that visit to Altoona, I told Mom for the first time about my dream as a child and for many years thereafter to build a time machine so I could go back and see Dad again. I explained some about the science and lasers involved, but I think she stopped listening. She looked deeply at me, with tears in her eyes.

"You look so much like your father," she finally said.

We hugged, and I cried with her.

Endnotes

1 (page xii) Danish physicist Niels Bohr (1885–1962) made essential contributions to the understanding of atomic structure as well as quantum mechanics. He received the Nobel Prize in Physics in 1922 for his work dealing with the structure of atoms. (The element bohrium is named in his honor.) During the Nazi occupation of Denmark in World War II Bohr escaped to Sweden and spent the last two years of the war in England and America, where he joined the Manhattan Project and helped build the atomic bomb. In his later years, he devoted his efforts to the peaceful application of atomic physics and to political problems arising from the development of atomic weapons. He died in Copenhagen.

2 (page xii) Hugh Everett III, born in Maryland in 1930 and raised in Washington, D.C., left physics a few years after receiving his Ph.D., discouraged at the lack of response to his theories from other physicists, including Niels Bohr, who was unimpressed with Everett's ideas. Everett soon went to work as a defense analyst and consultant for private industry, becoming a multimillionaire in the process. A chain-smoker and heavy drinker, Everett died suddenly in 1982 at age 51, apparently of a heart attack. His son, Mark "E" Everett, is the lead singer of the rock band Eels.

3 (page 22) H. G. Wells (1866–1946) was a biologist and his strong scientific background showed in his writings. His early novels—*The Time Machine,The Invisible Man,* and *The War of the Worlds*—billed as "scientific romances," invented

a number of themes now classic in science fiction. It is believed Wells was influenced by the popular writings of Jules Verne (1828–1905).

4 (page 23) Hendrik Antoon Lorentz (1878–1928) was born in the Netherlands. He won the 1902 Nobel Prize in Physics for his work on electromagnetic radiation, defining it so as to accurately explain the reflection and refraction of light. In 1904 he developed his Lorentz transformations, mathematical formulas that relate space and time measurements of one observer to those of a second observer moving relative to the first. As these formed the basis of Einstein's special theory of relativity (1905), Lorentz is considered one of the most prominent figures in the history of physics and the specific area of relativity. A man of immense personal charm, Lorentz endeared himself both to the leaders of his age and ordinary citizens who crossed his path.

5 (page 26) Georg Simon Ohm (1787–1854) was born and educated in Germany. After his breakthrough work in electricity, for which he used measuring and testing equipment of his own design, he had hoped to receive a prestigious university professorship. Instead, after his published findings were dismissed by the scientific community in 1827, he was forced to resign from his high-school teaching position and lived in poverty for a number of years. In 1841, he was finally recognized for the significance of his discovery and received a medal from the Royal Society of London. In 1849, five years before his death, he achieved his lifelong goal when he was appointed to a coveted position as professor of experimental physics at the University of Munich.

6 (page 36) In August 1955, in Money, Mississippi, a fourteen-year-old Negro youth from Chicago, Emmett Till, was dragged from his uncle's home by a group of white men. His head was crushed and he was shot and thrown into the Tallahatchie River. Till's murder was in retaliation for his going into a grocery store the previous day and whistling at the white wife of the store owner. Many consider the murder of Emmett Till to have been the spark that started the civil rights movement in America. Not until fifty years later, in June 2005, was there a conviction in the case, when Edgar Ray Killen, a former Ku Klux Klan member, was found guilty of manslaughter.

7 (page 42) Joseph John Thomson (1856–1940), an English physicist known by friends as J. J., is credited as the discoverer of the electron. His 1897 discovery caused a sensation in scientific circles, resulting in his winning a Nobel Prize for Physics in 1906. In perhaps the greatest father-son story of modern physics, his son, George Paget Thomson, was awarded a Nobel Prize for Physics in 1937 for proving that the electron is in fact a wave.

8 (page 47) Richard Feynman, a storyteller at heart, shared with Sin-Itiro

Tomonaga of Japan and Julian Schwinger the 1965 Nobel Prize in Physics for their "fundamental work in quantum electrodynamics, with deep-ploughing consequences for the physics of elementary particles." It turned out each man had done the same work independently. While Feynman's work as explained by his diagrams was famously easy to understand, Schwinger's overtly complex approach rendered his writings nearly indiscernible. When informed that Schwinger's work was mathematically equivalent to his own, Feynman exclaimed, "Now I have been translated into hieroglyphics."

9 (page 55) Erwin Wilhelm Mueller (1911–1977) was a German-born physicist who invented both the field emission microscope and the field ion microscope, the latter of which enabled him to be the first person ever to observe individual atoms. From 1952 until his death, he was a much-admired member of the Penn State Department of Physics, and received many recognitions, including (posthumously) the National Medal of Science from President Jimmy Carter in 1977.

10 (page 69) Einstein's new theories did not replace the Newtonian law of gravity, but simply completed an incomplete theory. When the speed of an object, such as a modern-day rocket, is much less than the speed of light, Einstein's relativity theory for motion becomes the same as Newton's theory of motion. Consequently, for weak gravitational fields and relatively slow speeds, Einstein and Newton are in agreement. In fact, Newtonian gravity was all that was needed for man to successfully go to the moon and back. For man's farther explorations of other galaxies, involving rockets flying nearer the speed of light, Einstein's theories will prevail.

11 (page 70) Newton and German mathematician G. W. Leibniz (1646–1716) independently discovered the basis for calculus, one of the major scientific breakthroughs of the modern era. Although Leibniz's seminal work on the subject was published first, Newton disclosed notes and other evidence showing that he had developed the same ideas twenty years earlier. For the last two decades of his life, Newton fought a heated legal battle with Leibniz over the authorship of calculus. It is now generally accepted that Newton developed calculus first, although it is still unclear when he would have shared it with the world had not Leibniz published first. In 1705, Newton became the first scientist to be knighted, and to this day is regarded as perhaps the greatest scientist of all time. Even Einstein ranked Newton number one; Einstein is usually ranked number two.

12 (page 73) Many years before Einstein, Austrian physicist Ludwig Boltzmann (1844–1906) established statistical mechanics as a means of showing that quantities like temperature were due to the average energy of motion of millions of molecules. Boltzmann's ideas were attacked by prominent physicists who disbelieved

the existence of molecules. Boltzmann was a man with a sensitive nature, and as a result of these rejections he suffered deep depression. He committed suicide by hanging in 1906 while on holiday in Italy. The previous year, Einstein had concluded his own work that would finally convince the scientific community of the existence of molecules, but in an era of slow communication word of Einstein's work did not reach Boltzmann in time.

13 (page 87) This Einstein-Lenard connection is ironic. In 1923, Lenard, addressing the German Physics Society, charged that "relativity is a Jewish fraud, which one could have suspected from the first . . . since its originator Einstein [is] a Jew." Lenard made this speech ten years before Hitler rose to power in Germany, an event that caused Einstein to flee his native country for the U.S. In the 1930s, Einstein helped many Jewish intellectuals living in danger in Germany to migrate safely. In the process, he became a leading Zionist. In 1948, he was offered the first presidency of the new state of Israel, which he declined on the grounds that he was not a politician.

14 (page 89) Harvard University professor Roy J. Glauber, eighty, was the core-cipient of the 2005 Nobel Prize in Physics for his work in developing a set of equations that accurately predict how photons behave in coherent light sources such as lasers. A student in the 1941 graduating class at Bronx High School of Science, Glauber attended only one year at Harvard before being recruited to work on the Manhattan Project, where he was assigned to calculating the critical mass of the bomb. He was only nineteen at the time. After working on the problem for two years, he went back to Harvard and obtained his bachelor's degree and his Ph.D.

15 (page 93) Michael Faraday (1791–1867) is often considered the greatest experimentalist in the history of science, even though he lacked a university education and knew only elementary mathematics. Refusing to believe Newton's premise that space is empty, Faraday, the son of a blacksmith, set up his experiments in a basement in his London flat. Largely responsible for electricity becoming a viable technology, he contributed significantly to the fields of electromagnetism and electrochemistry. He also invented the earliest form of what came to be known as the Bunsen burner, which is used universally in science laboratories as a convenient source of heat.

16 (page 93) Atomic clocks keep time better than other clocks, however, they are not radioactive, nor do they rely on atomic decay. Like ordinary clocks, they have an oscillating mass and a spring and use oscillations to keep track of passing time. The big difference is that the oscillation in an atomic clock is between the nucleus of an atom and the surrounding electrons, instead of between the

balance wheel and hairspring of a clockwork watch. Atomic clocks keep time better than the rotation of the earth or the movement of the stars. Without atomic clocks, GPS navigation would be impossible, the Internet would not synchronize, and the position of the planets would not be known with enough accuracy for space probes and landers to be launched or monitored.

17 (page 94) Ernest Rutherford (1871–1937), a native of New Zealand, is considered the father of nuclear physics. He won the Nobel Prize in Chemistry in 1908 for demonstrating that radioactivity is the spontaneous disintegration of atoms. During his investigation of radioactivity, he coined the terms alpha, beta, and gamma rays. Historians suggest that Rutherford is to the atom what Darwin is to evolution, Newton to mechanics, Faraday to electricity, and Einstein to relativity.

18 (page 95) I have since come to believe that it is a myth that their best work is done by the time physicists are in their twenties, although it is true that a number of outstanding contributions have been made by twentysomething physicists. However, due to the very nature of physics, which requires the accumulated acquisition of knowledge of the physical world, physicists often make groundbreaking contributions later in life. One need only mention Planck's quantum theory, Schroedinger's quantum mechanics, and Einstein's general theory of relativity; breakthrough work conducted by these pillars of physics at ages forty-two, forty-two, and thirty-seven respectively. And of course, each one of them contributed mightily to the discipline throughout their long lives. Einstein, in fact, was found working physics calculations on his deathbed in 1955, at age seventy-seven.

19 (page 111) John A. Wheeler was not the first to predict the existence of black holes. In 1939, American physicist J. Robert Oppenheimer, who would later direct the Manhattan Project and the building of the atomic bomb, and his graduate student, Hartland Snyder, published a paper in *Physical Review* entitled "On Continued Gravitational Collapse." This was the first prediction using Einstein's theory of general relativity regarding the final fate of a star sufficiently massive that light cannot escape it. Ironically, Wheeler initially disagreed with Oppenheimer's prediction, but later became an apostle of this final state, which he famously labeled a black hole.

20 (page 118) Joseph H. Taylor, Jr., born in 1941, and Russell A. Hulse, born in 1950, shared the 1993 Nobel Prize in Physics for "the discovery of a new type of pulsar, a discovery that has opened up new possibilities for the study of gravitation." Today, Taylor and Hulse, formerly teacher and student, respectively, are both professors of physics at Princeton University.

21 (page 126) Kip Thorne, born in Logan, Utah, in 1940, received his BS degrees from Caltech in 1962 and his Ph.D. from Princeton University in 1965. He returned to Caltech, where he is currently Feynman Professor of Theoretical Physics. His research is focused on Einstein's general theory of relativity and astrophysics, with emphasis on relativistic stars, black holes, and especially gravitational waves. More than forty physicists have received their Ph.D.s at Caltech under Thorne's personal mentorship. Thorne is known for his kind demeanor and rather modest bearing; he insists on being addressed by his first name and keeping in touch with his students even when he is on vacation.

22 (page 133) Karl Schwarzschild (1874–1916) was born in Frankfurt. A child prodigy of sorts, he had a scientific paper on orbits published when he was only sixteen. At the outbreak of World War I in 1914, he joined the German army despite being over forty years old, and served as an artillery officer. He died two years later, reportedly from an illness contracted while serving on the Russian front.

23 (page 136) Kris Larsen and I both consider our experience "partying with Hawking" to be the most unique at any of the many—and often staid—academic conferences we have attended. In fact, Kris, who today is a professor of physics and astronomy at Central Connecticut State University and director of the university's honors program, would describe the incident in the introduction to her first book, "Stephen Hawking: A Biography" (Greenwood Press, 2005). "According to Einstein, the speed of light is the ultimate speed limit of the universe," Kris wrote, "but somehow Ron managed to get dressed again and arrive at the party even faster."

24 (page 149) Richard Chase Tolman (1881–1948) was born in West Newton, Massachusetts. He received his Ph.D. from MIT in 1910 and was a professor of physical chemistry and mathematical physics and the dean of the graduate school at Caltech. He made many important contributions to statistical mechanics and cosmology, including the hypothesis of an oscillating universe (a closed-universe model in which the expansion of the universe slows and reverses, causing a collapse into a singularity which then explodes into a new universe, repeating the cycle). Tolman was a close friend of J. Robert Oppenheimer, who used mathematical techniques developed by Tolman to analyze the final state of collapsed stars, later to be known as black holes.

25 (page 157) Donald A. Glaser, born in 1926, has been a professor of physics and neurobiology in the Graduate School of the University of California (Berkeley) since 1989. He told reporters in Palm Coast, Florida, where he was vacationing at the time his Nobel Prize was announced, that winning the award

created a problem. "It makes it very hard to continue to do science. It's hard to discipline yourself to recognize that the next thing you do won't be as important."

26 (page 158) For a time, I considered the possibility that slowing down light might increase the gravitational frame dragging effect of the ring laser. Recent experiments by Lene Hau of Harvard University and Ronald Walsworth of the Harvard Smithsonian Center for Astrophysics had shown that light could be slowed down from 186,000 miles per second to only a few miles per hour. Slow light, however, turned out not to be helpful for my research.

27 (page 158) The special theory of relativity deals only with objects that have a constant and unchanging velocity, rather like a car that stays at a constant speed of sixty-five miles per hour. In contrast, the general theory is more general in nature (hence its name), dealing not only with objects that have a constant velocity but also ones that accelerate and decelerate.

28 (page 165) Philipp Lenard (1862–1947) was an Austrian-born physicist and winner of the Nobel Prize for Physics in 1905 for his research on cathode rays and the discovery of many of their properties. Lenard was a strong German nationalist and member of the National Socialist Party. During the Nazi regime, he was an outspoken proponent of the idea that Germany should rely on "Aryan physics" to conquer the world, and ignore what he believed to be the misleading and fallacious ideas of "Jewish physics," by which he meant chiefly the theories of Einstein.

29 (page 166) Photonics as a field began in 1960 with the invention of the laser. Thereafter, it evolved with the development of optical fibers as a medium for transmitting information using light beams, as well as the Erbium-doped fiber amplifier, inventions that formed the foundation for the telecommunications revolution of the 1990s and provided the infrastructure for the Internet.

30 (page 175) Charles H. Townes, born in 1915 in Greenville, South Carolina, received his Ph.D. from the California Institute of Technology. In 1964, he shared the Nobel Prize in Physics with two Russians, Nicolay G. Basov and Aleksandr M. Prokhorov, for fundamental work in the field of quantum electronics, which led to the development of masers (microwave amplification by stimulated emission of radiation) and lasers (light amplification by stimulated emission of radiation). Townes is presently at the University of California, Berkeley.

31 (page 182) C. Francis Everitt was born in Kent, England, in 1934. He received a Ph.D. in physics from the Imperial College, University of London, in 1959. As a research associate at the University of Pennsylvania, Everett conducted

research on liquid helium, which led him to become interested in using gyroscopes at very low temperatures as a possible means to test Einstein's general theory of relativity. Since 1962 he has been at Stanford as a professor of physics.

32 (page 183) Unlike an equatorial orbit, the plane of a polar orbit changes with the rotation of the earth. The equatorial plane rotates around itself, rather like putting a coin on its edge and rolling it across a table. The plane of a polar orbit is more like rotating (or flipping) the coin from one side to the other.

33 (page 184) On October 28, 2005, Francis Everitt spoke at UConn's physics colloquium during its Einstein Centennial celebration (the hundredth anniversary of Einstein's special theory of relativity). During a question and answer period, he was adamant about not discussing or speculating about the results of *Gravity Probe B* until the data has been analyzed. He estimated that results would be published sometime in 2007. Should Everitt and his team observe the frame dragging effect due to the rotating earth, it will support Einstein's general theory of relativity, as well as the foundation upon which my time travel work is based. If this turns out to be a negative experiment, in which Einstein's theory is disproved, it would be the major science story of my lifetime. What will not be proven by *Gravity Probe B* is whether frame dragging also occurs in the gravitational field of a circulating beam of light.

34 (page 186) The first contribution was made by New York composer and businessman David Zinn, whose father, William Zinn, was the musical arranger for Henry Mancini. Both Zinns came to UConn to meet with me, Chandra, and foundation director Frank Gifford for what felt like a high-energy summit conference between the arts and sciences.

35 (page 193) Present day space exploration is dominated by chemical rockets. These rockets have the advantage of great thrust. However, the velocities achieved make space travel to distant planets a very lengthy affair. There are other forms of rocket propulsion that are currently being considered that—while having little thrust—burn for long periods of time to provide a steady acceleration that can lead to very high velocities. An example of such an engine is the electrostatic ion propulsion drive. The rocket exhaust of such an engine consists of a beam of charged atoms or ions. Typically, a chemical rocket could make a flyby excursion of Pluto in forty-three years. In contrast, an ion rocket with a continuous acceleration as small as .0001 feet per second could make the journey in about three and a half years.

Index